BRITISH ISLES

Fig. 1 Definitions

BRITISH ISLES
England, Wales, Scotland, Ireland, Isle of Man, Channel Isles

UNITED KINGDOM
England, Wales, Scotland, Northern Ireland (Ulster), Isle of Man, Channel Isles

GREAT BRITAIN
England, Wales, Scotland—i.e. the mainland.

1a Name the islands numbered 1–5 on Fig. 1.
1b The area of Australia is approximately 3 million square miles. How many times is Australia larger than the British Isles?
1c Make a list of the 'countries' which make up the British Isles, placing them in order of size in square miles.

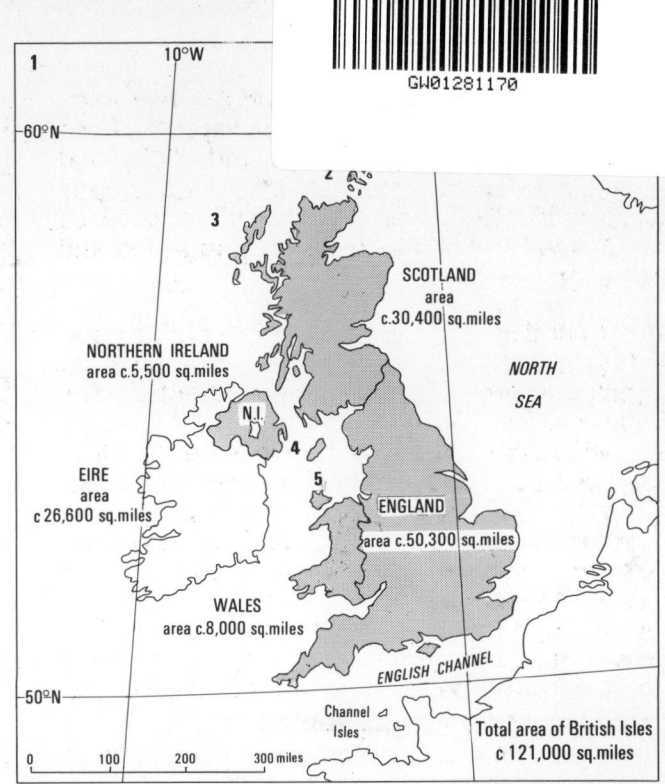

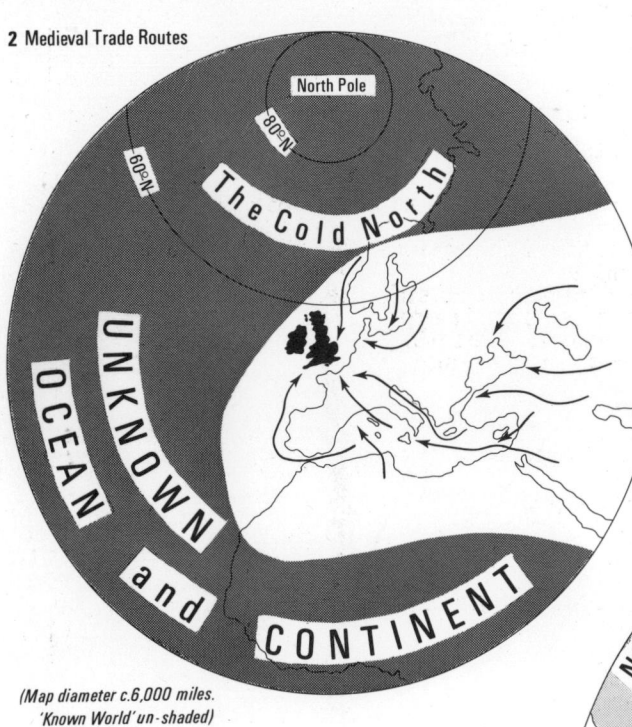

2 Medieval Trade Routes

(Map diameter c.6,000 miles. 'Known World' un-shaded)

3 English Commercial Expansion — C16th & C17th

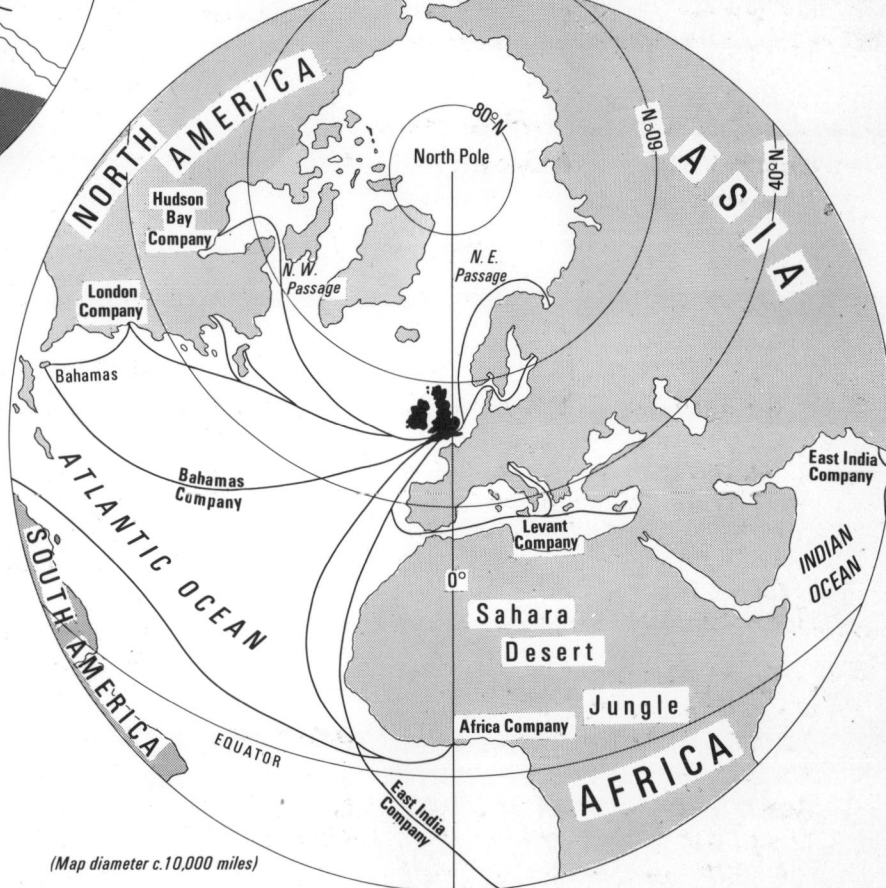

(Map diameter c.10,000 miles)

Figs. 2 and 3 Position

Fig. 2 shows the British Isles as lying on the edge of the 'known world' and trade routes of Medieval times. Most of the great continent of Africa and the whole of the Western World (North and South America) were then unknown to Europeans.

Fig. 3 shows that after the voyages of Columbus, and following explorations of the oceans and continents of the Atlantic, the British Isles became the middle of the 'known world' and its trade routes.

The opening up of the western hemisphere, which took place rapidly from the 16th century onwards, meant that the centre of gravity of World Commerce moved from the Mediterranean Sea area, in which it had rested for many centuries, to the Atlantic and its border-lands—especially to Western Europe and the British Isles.

2/3a Name important goods imported to UK by (i) Hudson Bay Co., (ii) Africa Company, (iii) East India Company.
2/3b Can you discover any important items which were exported from the British Isles during the 16th and 17th centuries?

2 BRITISH ISLES General Background

Figs. 4–11 This series of figures shows various invasions of the British Isles over many centuries. The last attempted invasion (by the Germans during World War II) failed.

Note the idea held by the Anglo-Saxons of the position of the British Isles right on the edge of the world they knew (east is at the top of the map).

4/11a Suggest some of the influences which the many invasions have had on the make-up, and even the character, of the British peoples.

4/11b Can you name any peaceful 'invasions' of Britain which have taken place within the last hundred years? (Think of immigrants.)

4/11c Why may we describe the British Isles as 'the final refuge of peoples pushing westward across Europe, until the opening up of the Western Hemisphere'?

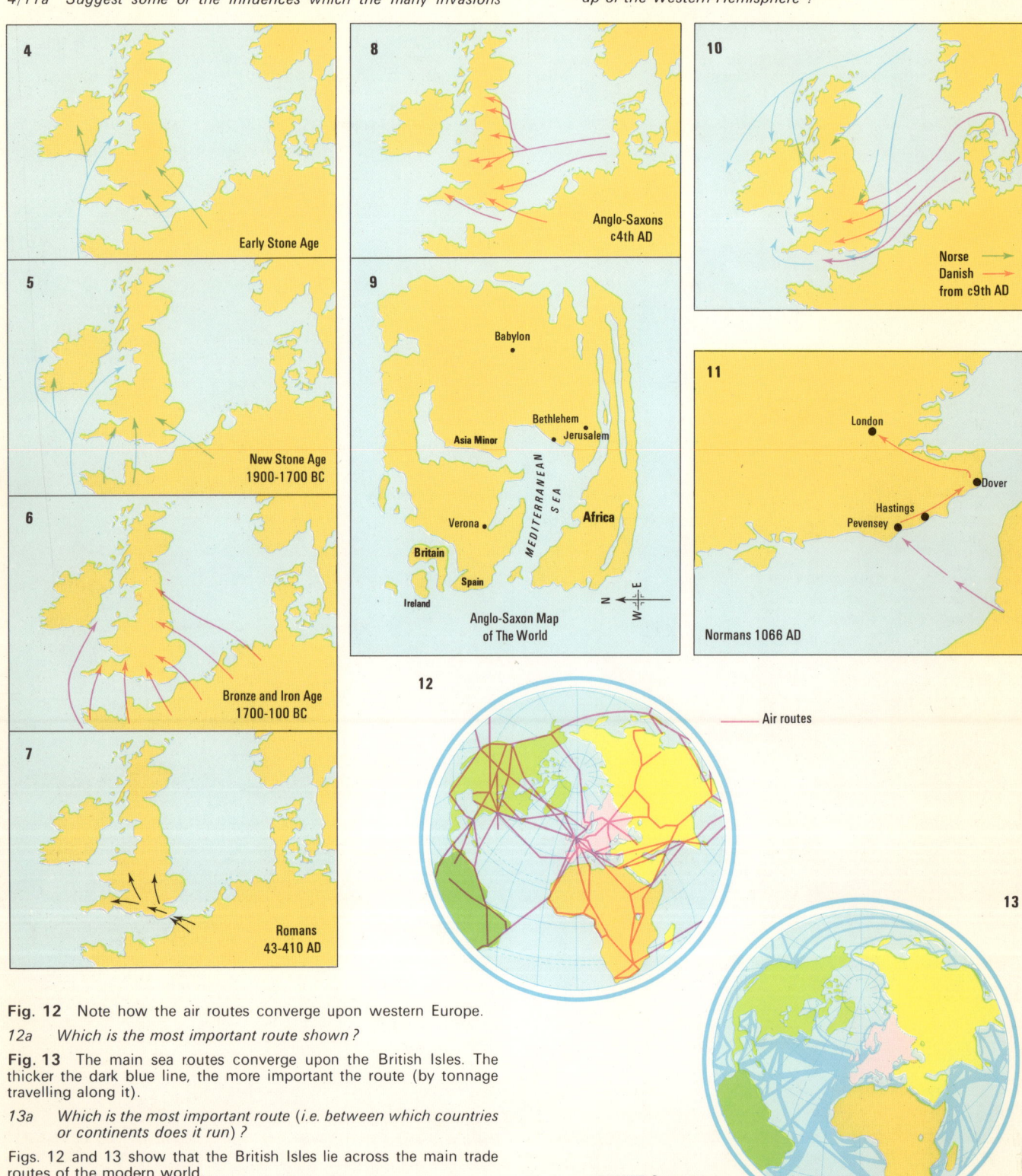

Fig. 12 Note how the air routes converge upon western Europe.

12a Which is the most important route shown?

Fig. 13 The main sea routes converge upon the British Isles. The thicker the dark blue line, the more important the route (by tonnage travelling along it).

13a Which is the most important route (i.e. between which countries or continents does it run)?

Figs. 12 and 13 show that the British Isles lie across the main trade routes of the modern world.

BRITISH ISLES Relief and Structure

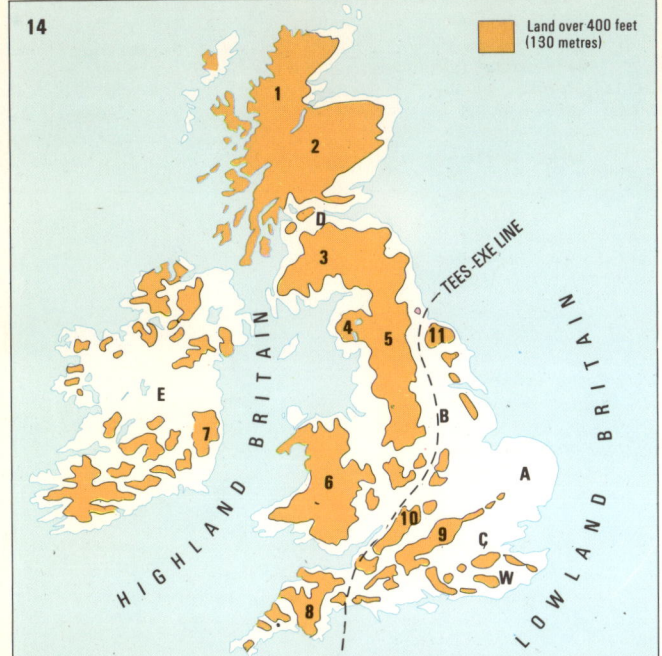

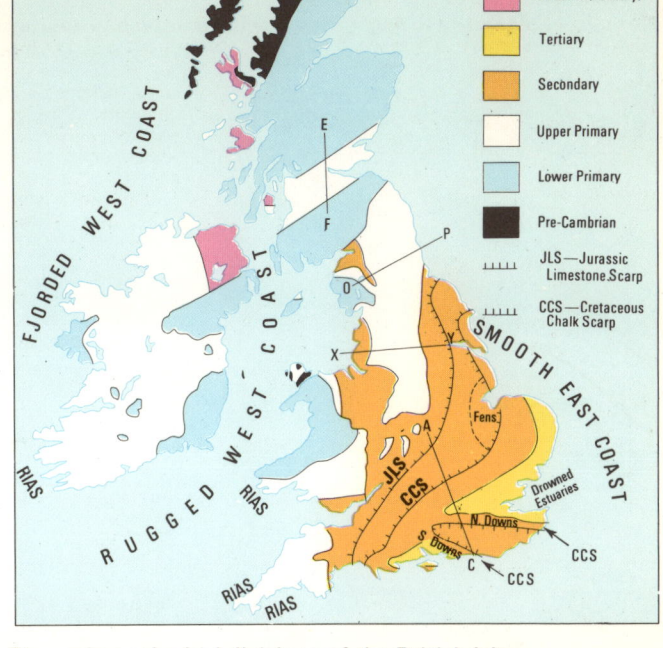

Fig. 14 Relief is the result of the interaction of geological structure, the nature of the rocks, and the type, stage and history of the erosion and deposition processes (the work of the 'agents of denudation'—wind, water, ice, waves, weather). Fig. 14 shows two relief sections of the British Isles—Highland Britain and Lowland Britain. These divisions can also be seen on Figs. 15 and 16. The Jurassic Limestone scarp is a convenient structural feature which we may use to divide these two Britains.

14a Identify upland areas marked 1–11.
14b Identify lowland areas marked A–E.
14c Identify area marked W.
14d Describe, with the aid of sketches, the relief of your own area.

Fig. 15 The broadly triangular shape of the British Isles is basically the result of THREE folding movements of widely separated periods in our geological history:
  (i) the Caledonian (structure running roughly NE-SW)
 (ii) the Hercynian (structure running roughly E-W, but having a N-S trend in the Pennines).
(iii) the Alpine, which directly affected only south-east England, but whose disturbance allowed volcanic activity to take place in Scotland and Northern Ireland.

15a Find the areas affected by each of these movements on Fig. 15.
15b Which are (geologically) the oldest parts of the British Isles, and which the youngest?

Fig. 16 The main geological divisions of the British Isles

16a Examine Figs. 14, 15 and 16 together. Find the Jurassic limestone scarp and the chalk scarps, and name the **limestone** Cotswolds, the Northampton Uplands, the Lincolnshire Edge, the Yorkshire Moors; and also the **chalk** Yorkshire and Lincolnshire Wolds, the East Anglian Heights, the Chilterns, Salisbury Plain, and the North and South Downs.
16b What is meant by (i) ria, (ii) fjord?

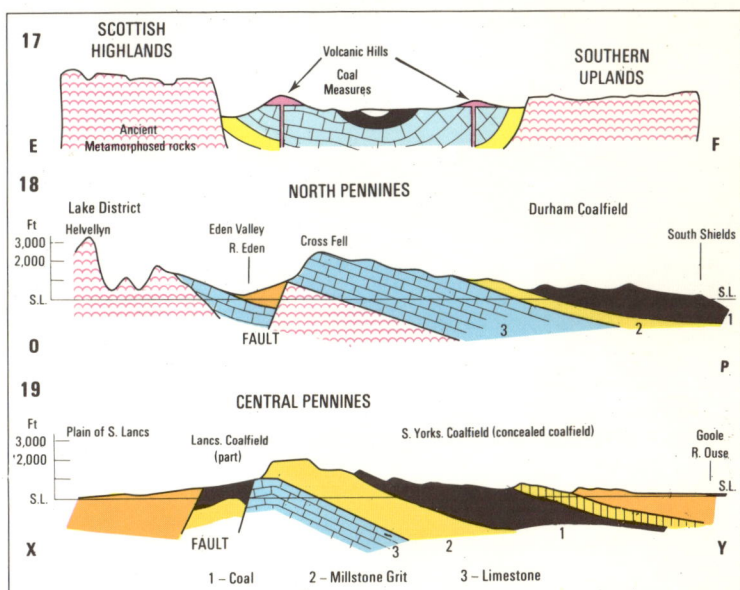

Figs. 17–20 Examine each section line on Fig. 16 and the corresponding cross-section (Figs. 17–20).

17a What is the main structural feature of the Midland Valley of Scotland?
18a What type of fault is shown in Fig. 18?
19a Which part of the coalfield of S. Yorkshire has the deepest mines?
20a How many scarps are shown in Fig. 20?
20b Try to discover the geology of your own area; collect and name rock specimens.

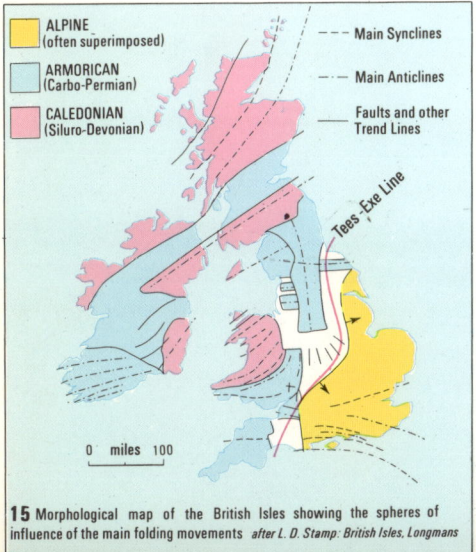

15 Morphological map of the British Isles showing the spheres of influence of the main folding movements *after L. D. Stamp: British Isles, Longmans*

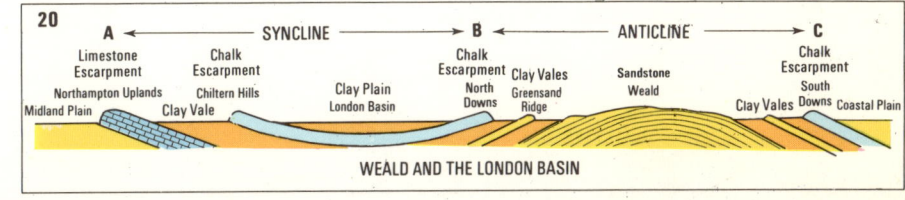

4 BRITISH ISLES Relief, Structure, Glaciation

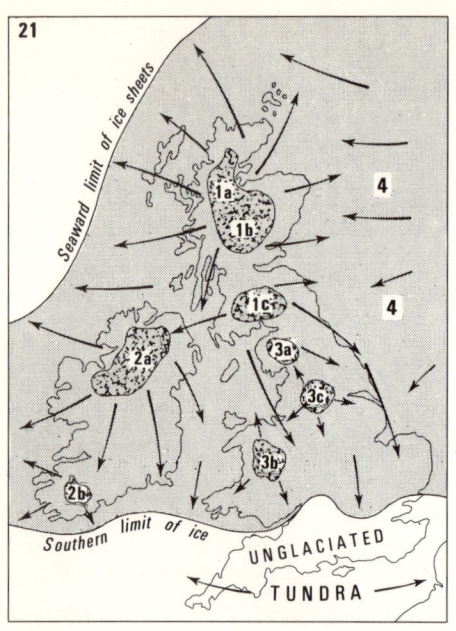

Fig. 21 The whole of the British Isles experienced, in some form or other, the influences of the Quaternary Ice Age—whose ice sheets melted from these islands only some 10,000 years ago. Most was actually buried beneath the moving ice and suffered great change in surface features by direct ice erosion and deposition. The parts which were not directly covered by ice had a tundra type of climate, with much frost action. The surface features were also greatly affected by the action of running melt-waters in immediate post-glacial times, and a general post-glacial rise in sea-level caused the drowning of much of the coast.

21a The main centres of ice-dispersion are numbered on Fig. 21. Suggest a name for each.

21b What have these centres in common (examine Fig. 16)? Suggest reasons for the common factor.

21c Find out and write about the characteristics and present locations of tundra climates.

21d How has the drowning of parts of the coast helped shipping?

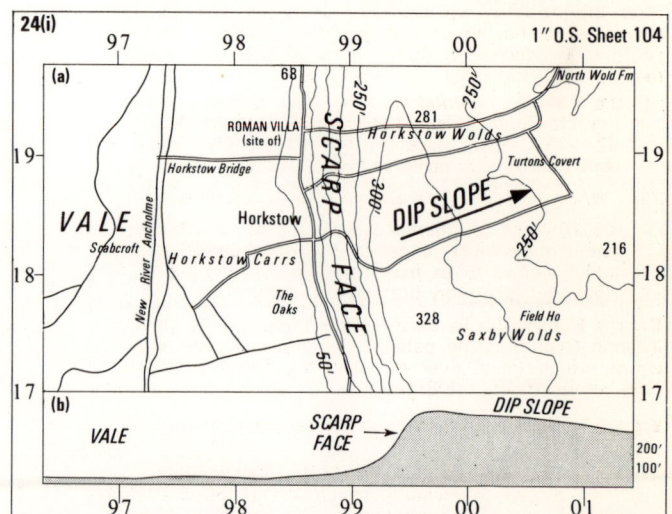

Fig. 22 Location of glacial deposits

22a Which parts of Britain have most glacial deposits? How would you describe the general relief of these areas?

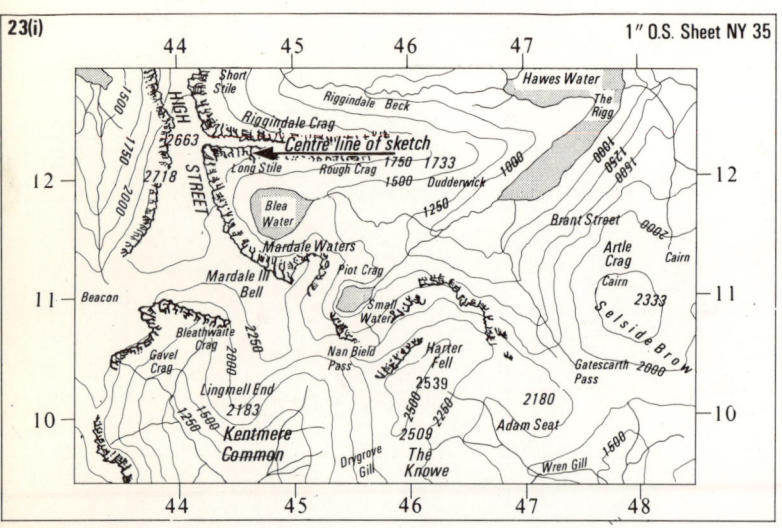

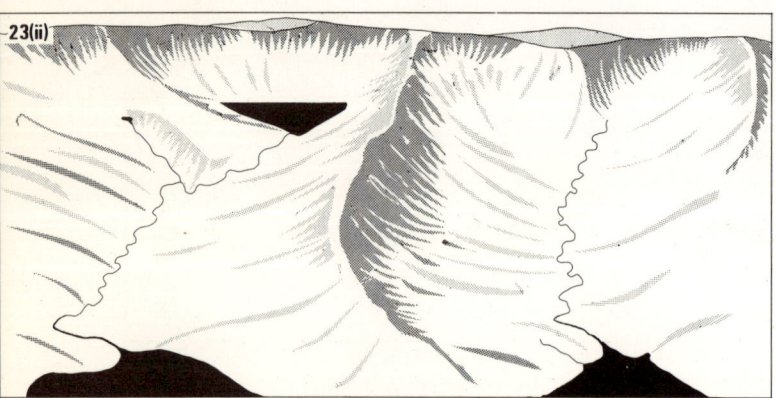

Fig. 23 The section of O.S. map and the sketch both show the Blea Water area of the Lake District—typical of much of Highland Britain.

23a Identify on the map the features on the sketch.

23b What is the total area of the map in sq. kilometres?

23c In which direction was the artist looking when he sketched the area?

23d Find and describe the following glacial features of the area: corrie (cirque or cwm); arête; glacial trough.

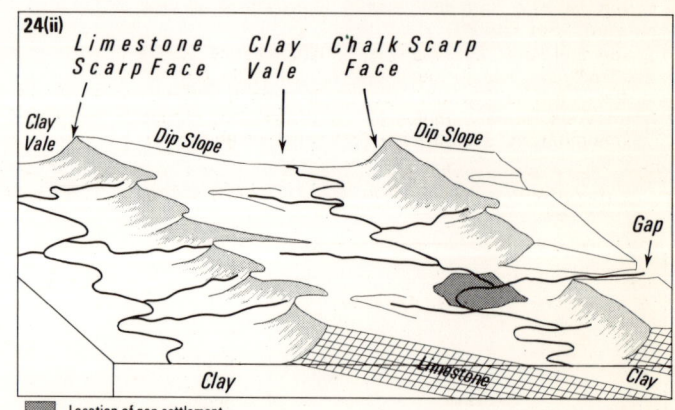

Fig. 24 The map shows a small section of a chalk scarp. The block-diagram shows a typical scarp-and-vale formation of lowland Britain.

24a In which direction is the scarp (on the map) facing?

24b What are the advantages of the location of the gap settlement shown on the block-diagram?

24c Is the scarp-face on the map dissected?

24d Find two examples of gap settlements in England.

BRITISH ISLES Climate

The climate of any place is the average of the weather it experiences during the year (usually worked out over a 35-year period).
In the British Isles the main controls of climate are:
1. latitudinal position
2. location in relation to Eurasia and the Atlantic
3. general and local relief.

1. LATITUDINAL POSITION

This puts the islands in the *cool temperate climatic belt* and ensures that there are considerable changes in the angles of slope of incoming sun's rays during the year, notable between June 21st and Dec. 21st. Thus we are provided with winter, summer, spring and autumn. See Fig. 25.

Fig. 25 The angle of the sun's rays

25a At which lines of latitude is the sun directly overhead on Dec. 21st, June 21st, March 21st, September 21st?

2. LOCATION

A The British Isles lie on the western edge of a very large land-mass (Eurasia) which has for the most part cool or cold winters and warm to hot summers, and from which winds, with some easterly component and varying temperatures and humidities, are received.

Figs. 26 and 27 Two main air masses (an air mass is a body of air having distinctive characteristics of temperature and moisture) come to the British Isles from the east and south-east:
(i) the *Polar Continental air mass* (Pc) in winter—see Fig. 26. Characteristics: low temperatures (5–8°C./42–45°F.), hard black frosts, cold easterly winds, clear blue skies or layers of low stratus.

26a When did you last experience this kind of weather?

(ii) the *Tropical Continental air mass* (Tc), which we rather occasionally experience during summer and which comes to us from the south or south-east. Characteristics: very hot and dry; hazy skies.

B The British Isles lie on the eastern edge of the great Atlantic Ocean, in the path of the *prevailing westerly winds* which travel over the warming *North Atlantic Drift* before striking the British Isles.

Figs. 28 and 29 Two main air masses come to Britain from the seas:
(i) the warm *Tropical Maritime* (Tm). Characteristics: mild, muggy, enervating; much stratus cloud and hill fog; sea fog in summer; drizzle and relief rain; high relative humidity; dismal. Experienced all year round. See Fig. 28.
(ii) the cool *Polar Maritime* (Pm). Characteristics: winds usually have a northerly component; cold but not freezing; raw in winter; cool and bracing in summer; showers; large cumulus or cumulo-nimbus clouds commonly set in a blue sky. Also experienced all year.

29a Keep a log describing the climate in your own area (temperature, precipitation, clouds, winds) during the year.

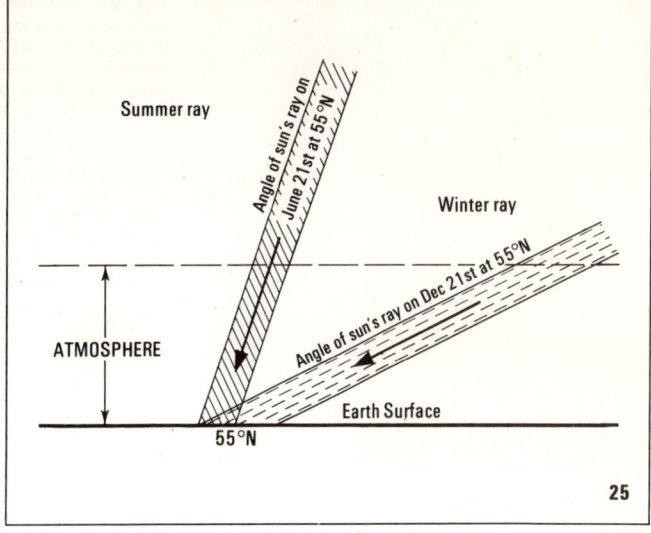

Notes 1. Both 'parcels' of the sun's rays are of the same width and power outside the atmosphere.
2. The June ray has a thinner band of atmosphere to penetrate and heats a smaller area of earth surface. Hence less heat is lost, and temperatures in the British Isles are higher in summer than in winter.

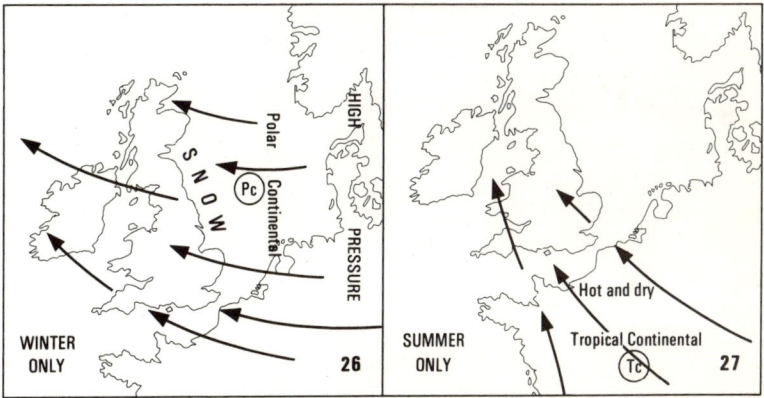

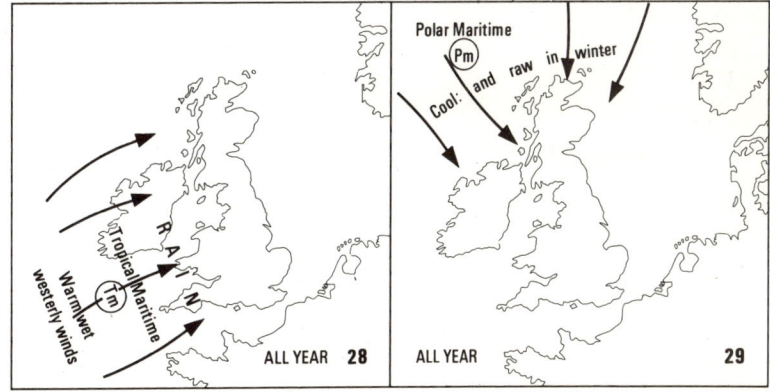

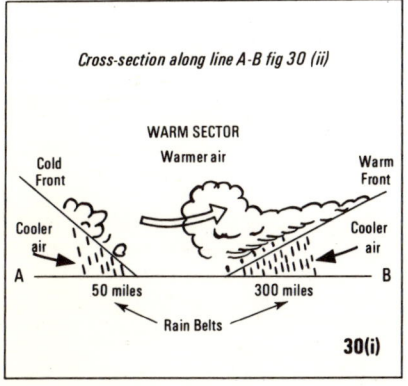

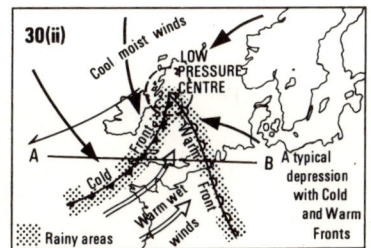

Fig. 30 Fronts

When air masses of different characteristics meet, 'fronts' are formed between them. The rising of the damp air along the fronts (warm or cold) may cause sufficient cooling for clouds and rain to form. Areas of *low pressure*, in which air masses may meet and such fronts be formed, may be experienced at any season. Some 40 low pressures, or *depressions*, cross the British Isles each year, and account for much of our rain.
Fig. 30(i) shows a cross-section along line A-B on Fig. 30 (ii).

BRITISH ISLES Climate

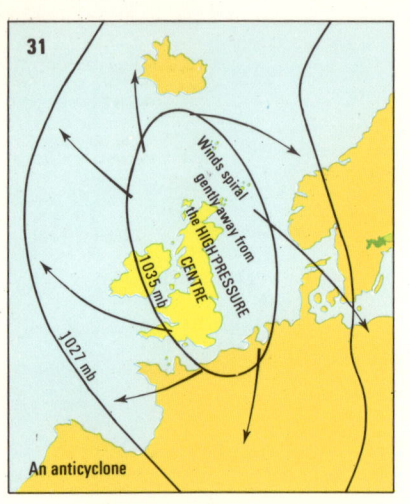
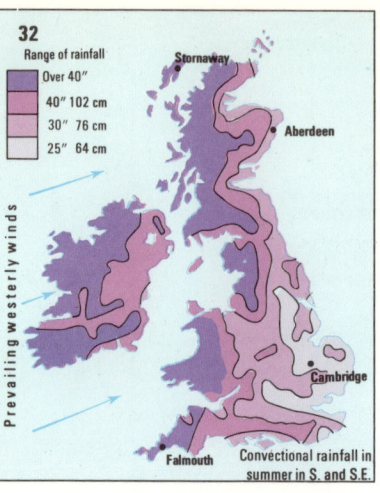
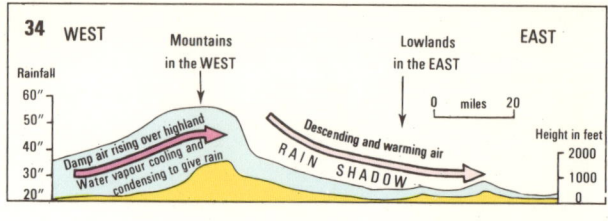
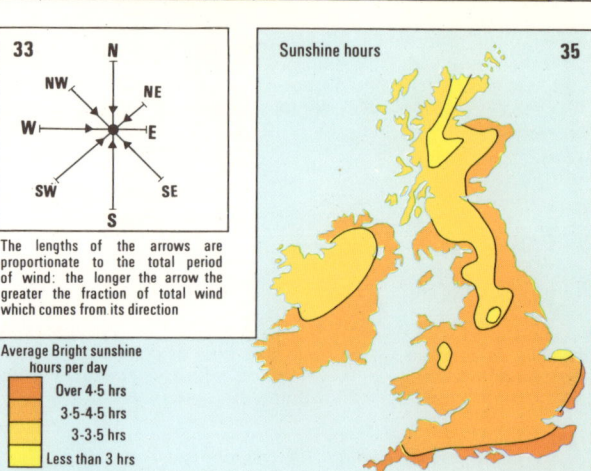

Fig. 31 Anticyclones are slow-moving cells of high atmospheric pressure. They bring calm conditions. The high pressure excludes outside weather elements and thus allows high temperatures to develop in summer and cold, frosty, clear weather in winter. The British Isles have only about a dozen anticyclones per year on average.

31a What is the direction of wind circulation around the high pressure?
31b Are the winds blowing into or out of the centre of the anticyclone?

3. GENERAL AND LOCAL RELIEF

Fig. 32 Temperatures decrease by about 3°F. for each increase of 1000 feet of height (0.6°C. per 300 metres). Thus upland areas are usually cooler than lowlands. Moreover, the concentration of high land on the *west* means that warm damp westerly airstreams are forced to rise there and to deposit much of their moisture as relief rain on the western side of the British Isles. See also Fig. 34.

32a What is the average annual rainfall in your area?
32b How many inches of snow usually fall there per year?
32c What influence does the relief of your area have upon local temperatures and rainfall?

Fig. 33 Wind rose for Stornoway

33a What is the prevailing wind of the British Isles shown by Fig. 33? Put the wind directions in order according to the amount received from each quarter. Find Stornoway on Fig. 32.

Fig. 34 Cross-section

34a Write a description of what is shown by Fig. 34.

Fig. 35 Sunshine hours

35a Describe the relationship between Figs. 32 and 35.

Fig. 36 Winter isotherms

36a In which general direction do the isotherms run across the British Isles during winter? Give reasons for the direction.

Fig. 37 Summer isotherms

37a In which general direction do the isotherms cross the British Isles during summer? Give reasons for this, and explain why the isotherms curve to the south over the seas.

Figs. 38–41 Climographs

38/41a Note the positions of each of the places for which climographs are given on Fig. 32.

38/41b Explain
   (i) why the winter temperatures at Cambridge are lower than those of Stornoway, which lies further north,
   (ii) why Stornoway and Falmouth have more rain than the other two places,
   (iii) why neither temperatures nor rainfall cause serious difficulties for plant growth during the year at Falmouth.

38/41c How many months below 43°F./6°C. are there at each place? State what is important about this temperature.

38/41d Draw a simple sketch map of the British Isles and mark on it the isotherm for 60°F./15.5°C. (in red) and the isotherm for 40°F./5°C. (in blue). This will divide the country into four quarters (quadrants). Using all the information given about climate here, describe briefly on your map *the main characteristics of the climate of each quadrant* (winter and summer temperatures, and precipitation).

38/41e List all the factors which influence the British Isles climate.

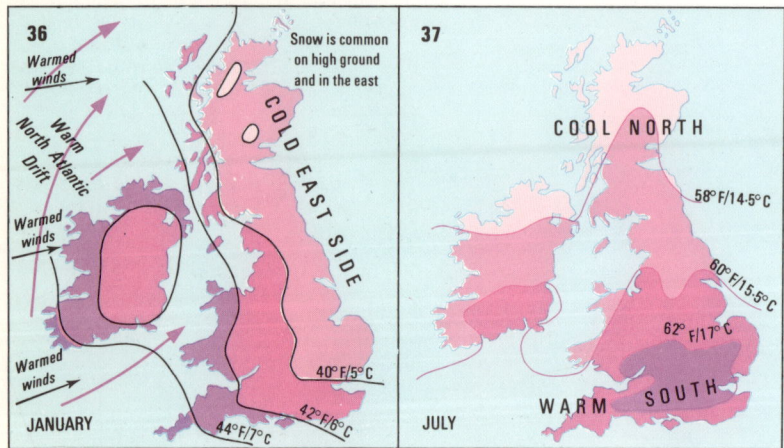
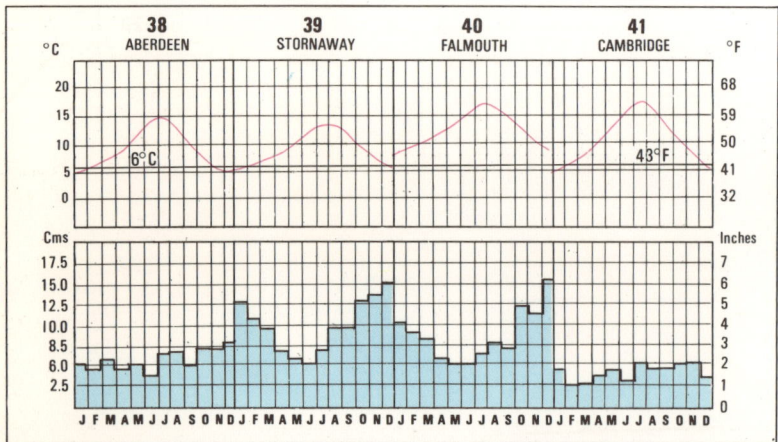

BRITISH ISLES Population 7

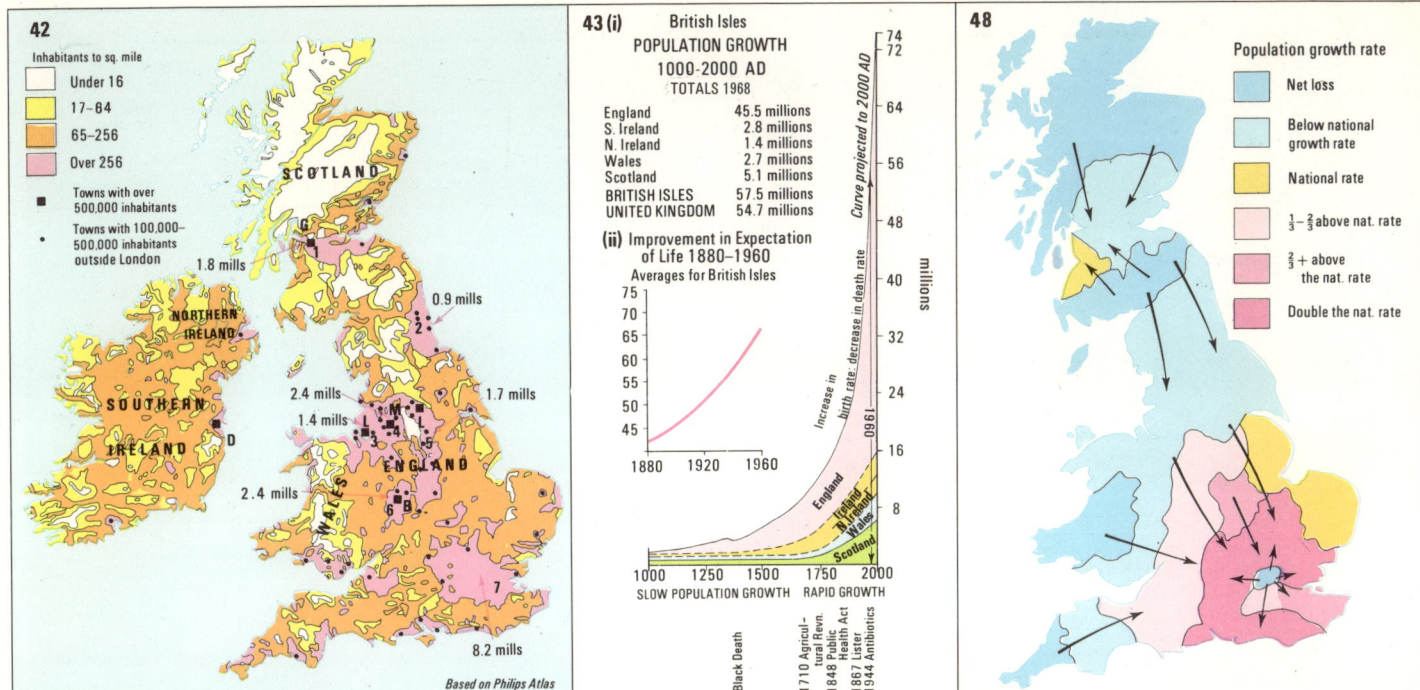

Fig. 42 Population distribution

42a Examine Fig. 42 carefully together with an atlas map of the British Isles. Then
(i) name the cities with populations over 500,000, for which the first letters have been marked on the figure,
(ii) suggest names for the conurbations numbered 1–7 on the figure (e.g. No. 1 might be called the Clydeside region).

Fig. 43 Population growth

43a Suggest reasons for the rapid increase in population since 1750.

Fig. 48 In recent years there has been a move away from the coalfields: new forms of power (thermal, hydro and atomic electricity; oil, and natural gas) have released industry from the necessity of a coalfield location. The big towns of the south-east have developed much light industry. The high opportunities for well-paid employment, and the presence of London, act as powerful magnets drawing people to this part of the British Isles. This movement is often called 'the drift to the south-east', and it is illustrated in Fig. 48.

48a Why should people want to live near to London?
48b Has population increased within London itself? If not, why not? (Think of the high prices of land and goods in general, and government restrictions.)
48c Where do the majority of the people who work in London actually live?
48d What influence does the drift to the south-east have on house and property prices in general in the suburban areas of the south-east?
48e Why is the City of London 'dead' at night?
48f Suggest reasons for the decline in population in Cornwall, Wales and Scotland. (Think of relief, climate, location, amenities, the availability of good jobs.)

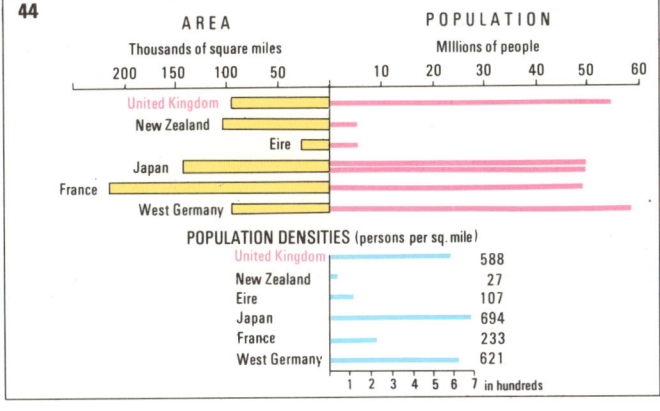

Fig. 44 Population density

44a Examine the figure and then compare and contrast the area, population figures and densities of the UK with those of New Zealand, of France, and of any one other of the countries listed.

Figs. 45–47 These three maps show the areas of high population density (over 400 people per square mile) at three points in history.

45/47a Describe the changes which you notice between 1851 and 1961. (The high density areas in 1851 were almost all on the coalfields. See Figs. 42 and 103.
45/47b Contrast the urban percentages for Ireland and Great Britain (i) in 1851, (ii) in 1961

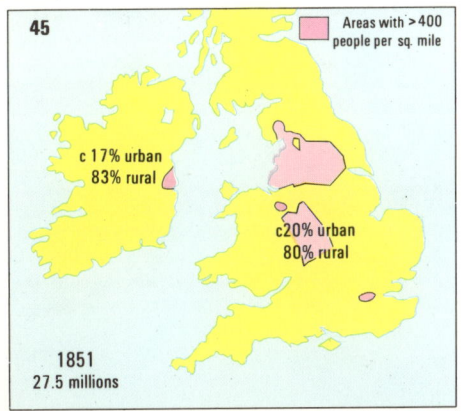
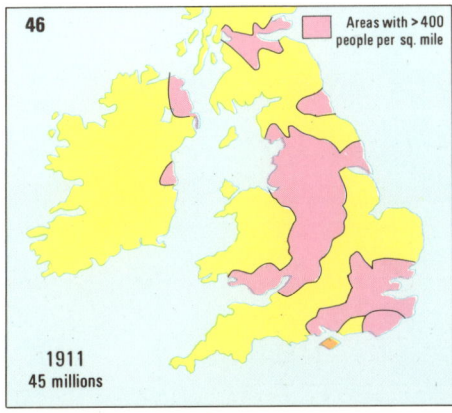
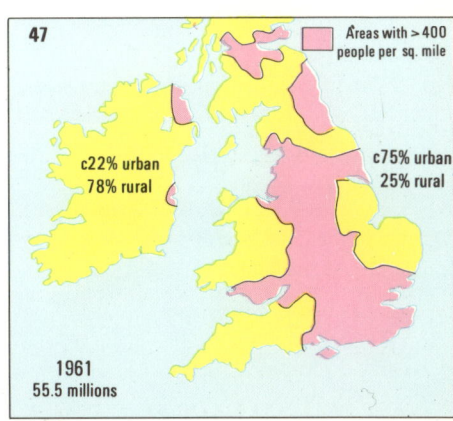

8 BRITISH ISLES Population and employment

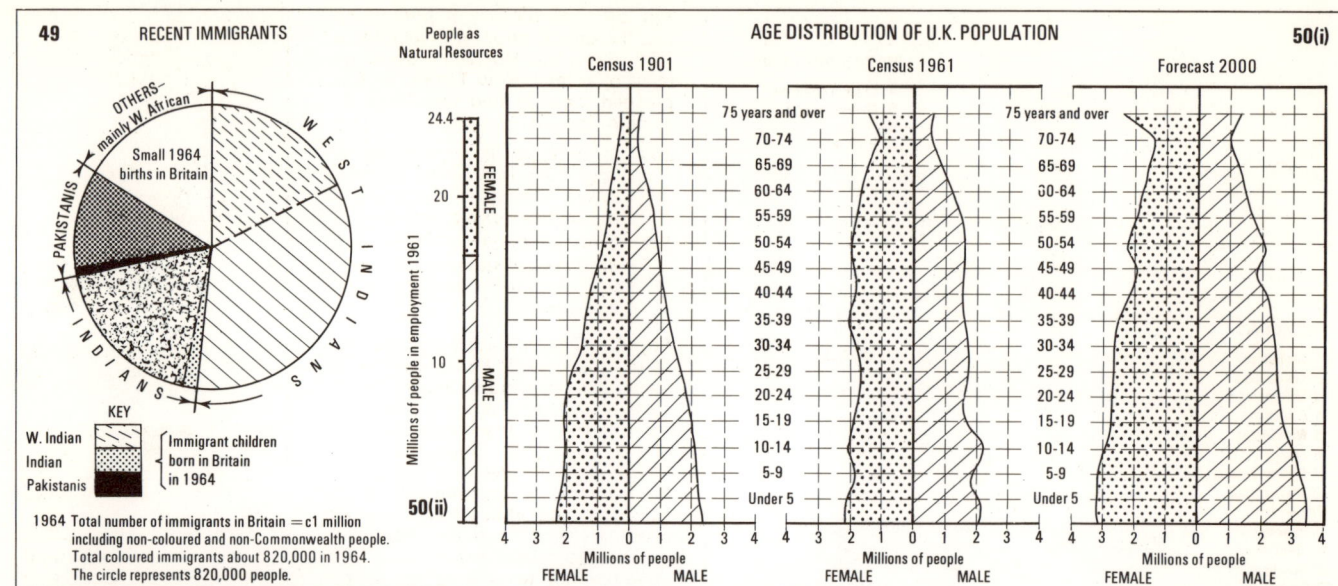

Population increase takes place when the number of 'new people' (i.e. new births plus new immigrants) exceeds those departing (through emigration and death). In recent years there has been a considerable flow of peoples into the British Isles—notably from Commonwealth countries—and especially into certain industrial towns where work is available, such as Birmingham, Bradford, London and Nottingham. Here they tend to congregate in groups in the same parts of the towns.

There is also an outflow from the British Isles, especially to Canada, Australia and New Zealand—usually of young, skilled people—and there is also a flow of fully qualified professional people and scientists to highly paid posts in the USA (the 'brain drain').

Fig. 49 Immigrants

49a Place the immigrants in order of importance, judged by the total numbers in each group.

49b To which group were the greatest number of children born in Britain in 1964? How old will these children be now?

Fig. 50 In the process of production (the provision of goods and services) the main factors are:

(i) the land surface—including all the natural elements above and the minerals below,
(ii) the capital equipment available and the rate of investment in new capital and research,
(iii) motive power to drive the machines,
(iv) the *work-force*, i.e. the working population.

The *effectiveness* of the work-force itself depends upon many factors, including the total number of workers, its make-up (sex ratio, and age groups), its quality (intelligence and training), its attitudes to life and work and towards the idea of 'women's work', its distribution between various occupations, and its 'mobility' or ease of movement both between different types of work and about the country from place to place also.

50a Describe the main differences between the population pyramids for 1901 and 1961, and between the 1961 pyramid and that projected for the year 2000.

50b Why is there an increase in the older age groups in 1961 and 2000?

50c Why is it important to have some idea of the age make-up of the population? (Think of size of work-force, retirement ages, demand for services and goods.)

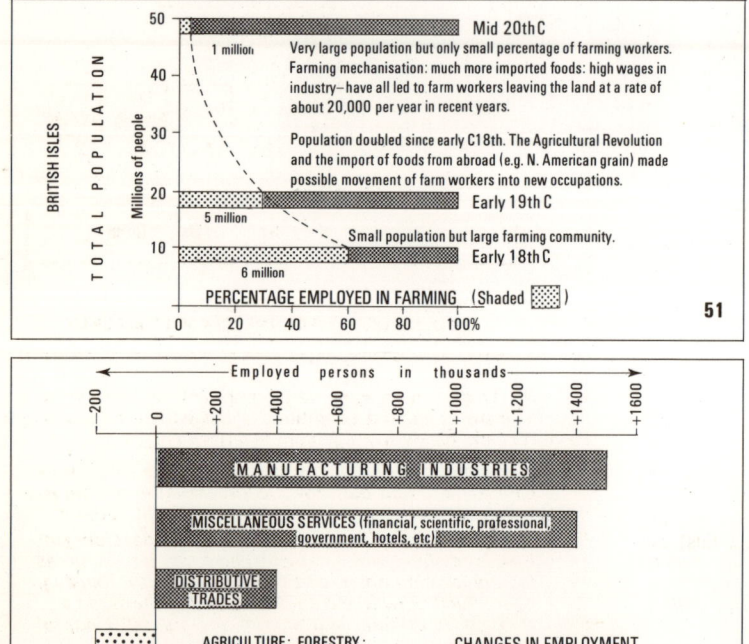

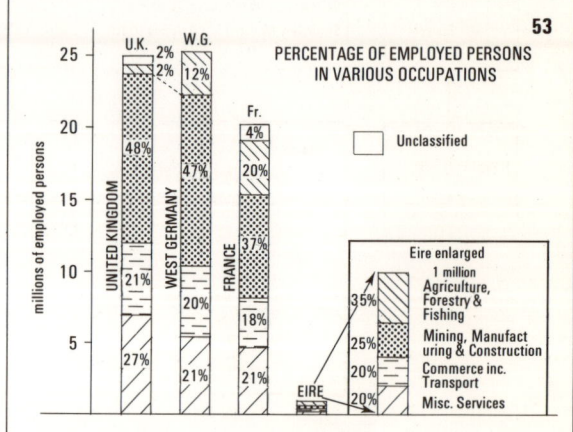

Figs. 51, 52 and 53 Employment

51a What is the main point illustrated in Fig. 51, and how is it further illustrated in Fig. 52?

51b How many were in agricultural work (i) in early 18th C., (ii) in early 19th C., (iii) in mid-20th C.?

52a What is shown by Fig. 52?

53a Place the countries in order according to the percentages of totals devoted to (i) agriculture, forestry and fishing, (ii) mining, manufacturing and construction.

BRITISH ISLES Population and employment

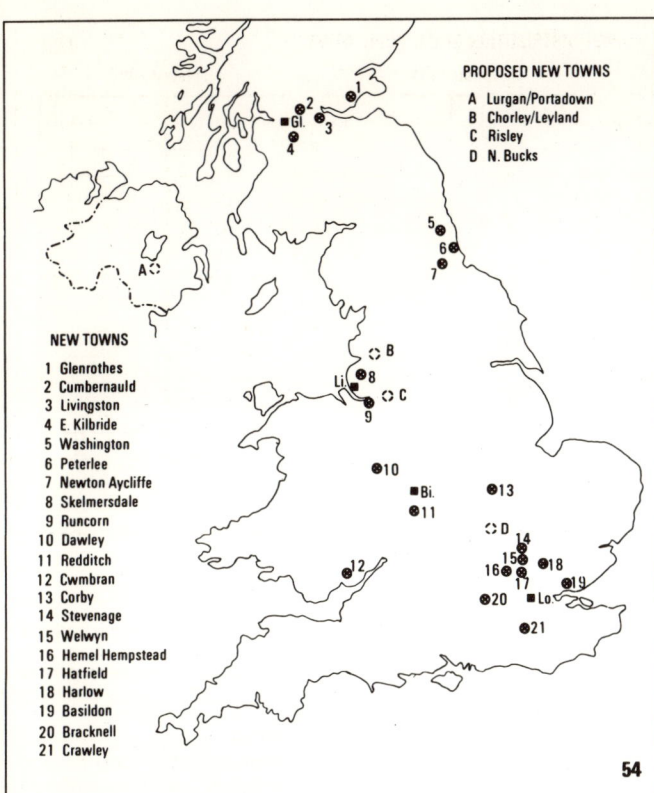

PROPOSED NEW TOWNS
A Lurgan/Portadown
B Chorley/Leyland
C Risley
D N. Bucks

NEW TOWNS
1 Glenrothes
2 Cumbernauld
3 Livingston
4 E. Kilbride
5 Washington
6 Peterlee
7 Newton Aycliffe
8 Skelmersdale
9 Runcorn
10 Dawley
11 Redditch
12 Cwmbran
13 Corby
14 Stevenage
15 Welwyn
16 Hemel Hempstead
17 Hatfield
18 Harlow
19 Basildon
20 Bracknell
21 Crawley

Fig. 54 The growth of population and the deterioration of dwellings in 'old' towns has led to the development of several 'New Towns' and to planned clearing and rebuilding of some of the old.

54a In what main respect would you think the layout of new towns differs markedly from the patterns of old towns?

54b Why are blocks of flats often built several storeys high nowadays? (*Think of land shortages and costs.*)

Fig. 55 The drift to the south has been caused partly by the decline of once very important industries in parts of northern, western and central Britain—e.g. the cotton industry of Lancashire, the ship-building of the Clyde and Tyne, coal-mining in South Wales. The closing of mills, shipyards and mines led to heavy unemployment in these areas. In order to help those unemployed to find work in their own areas (and therefore not to drift to an already heavily populated south-east England), the Government set up Development Areas (shaded black on Fig. 55) in which inducements such as cheap loans for building, equipment and raw materials have been made available for those wishing to set up new industries.

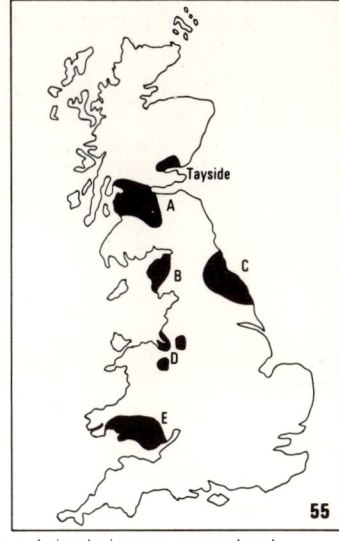

55a What important factor of production was already available in the Development Areas?

55b Suggest names for the areas marked on the figure.

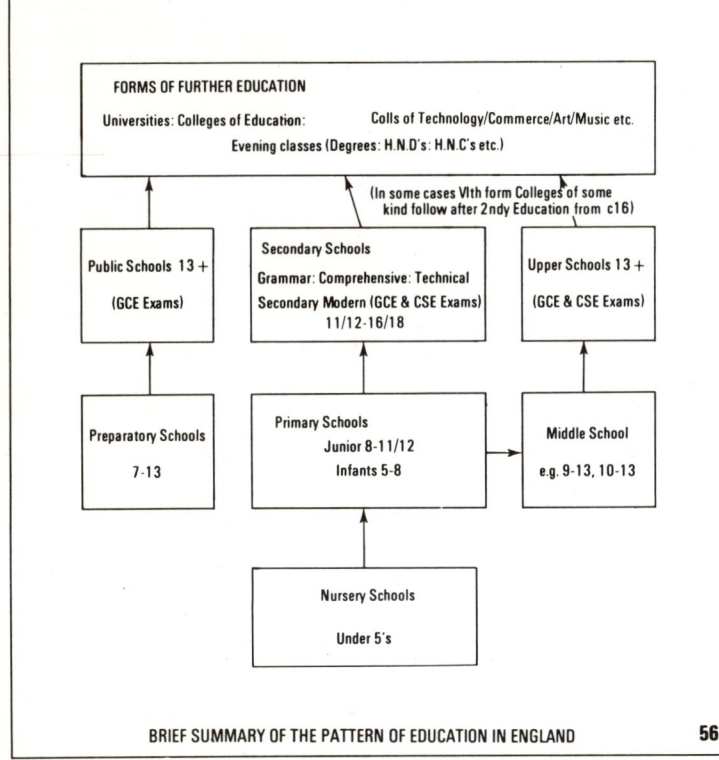

BRIEF SUMMARY OF THE PATTERN OF EDUCATION IN ENGLAND

Fig. 56 The development of any region depends, as we have seen, not only on the obvious physical factors (relief, climate, location, minerals) but also on the work-force (this term includes management) and its general efficiency and quality. These are very much influenced by *education standards,* by the organisation of labour into *Trade Unions,* and by *Government policies and influences.*

56a Which of the 'education routes' shown on Fig. 56 are you following or do you intend to follow in the future?

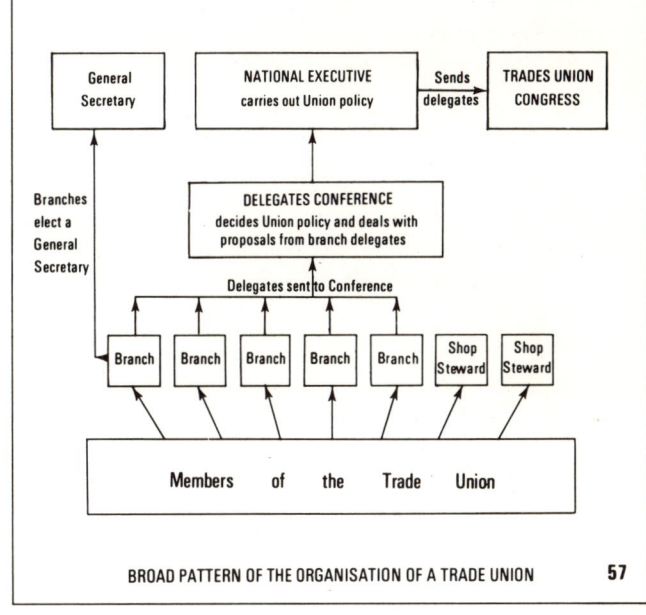

BROAD PATTERN OF THE ORGANISATION OF A TRADE UNION

Fig. 57 Trade Unions are very important in helping to determine the terms and conditions under which the labour force (usually below management level) works.

57a If any member of your family is a member of a Trade Union, try to find out how it is organised and what its various functions are, and write an account about it.

57b The Government influences economic development (and therefore very many aspects of life, such as factory location and growth, transport policy, housing, agricultural growth and land use) through taxation and through its attitude and actions towards all kinds of development.
What is meant by (*i*) the C.B.I., (*ii*) E.F.T.A., (*iii*) N.E.D.C.?

10 BRITISH ISLES Land Use

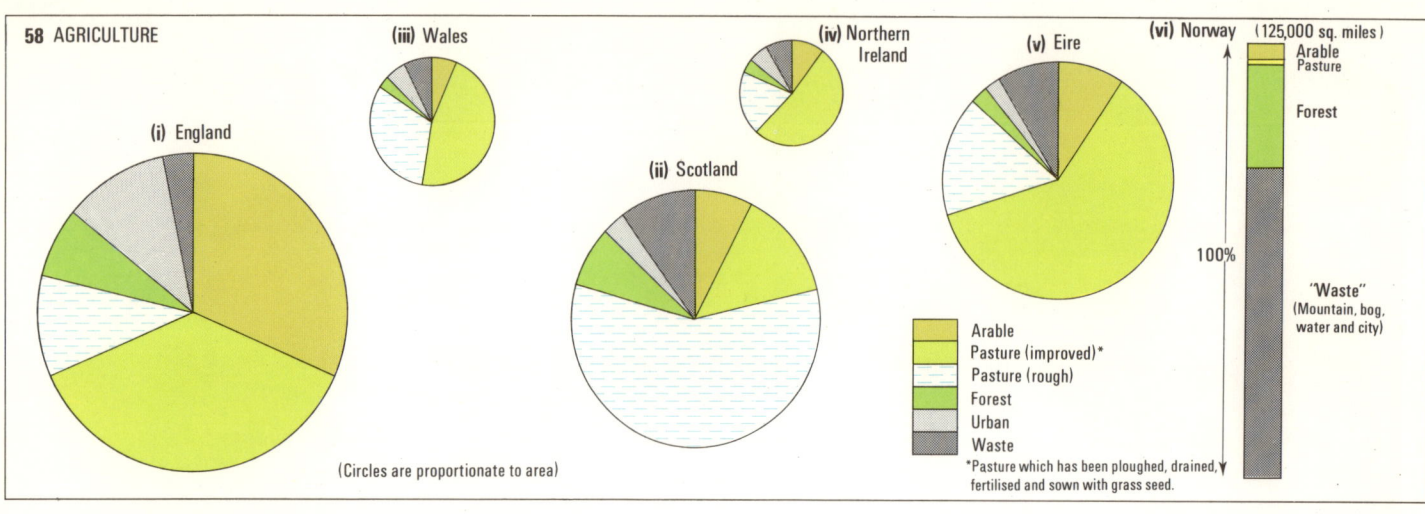

Fig. 58 The pie graphs illustrate Land Use in the British Isles. The bar graph shows the same for Norway.

58a Why is there so much rough pasture in Scotland and Wales?

58b Which country has the most arable land, and which the least?

58c What climatic factor has encouraged the large percentage of permanent pasture in Eire?

58d Why is there so much 'waste' in Norway?

58e The Land Use of which country of the British Isles is most like that of Norway? Suggest reasons for the similarities.

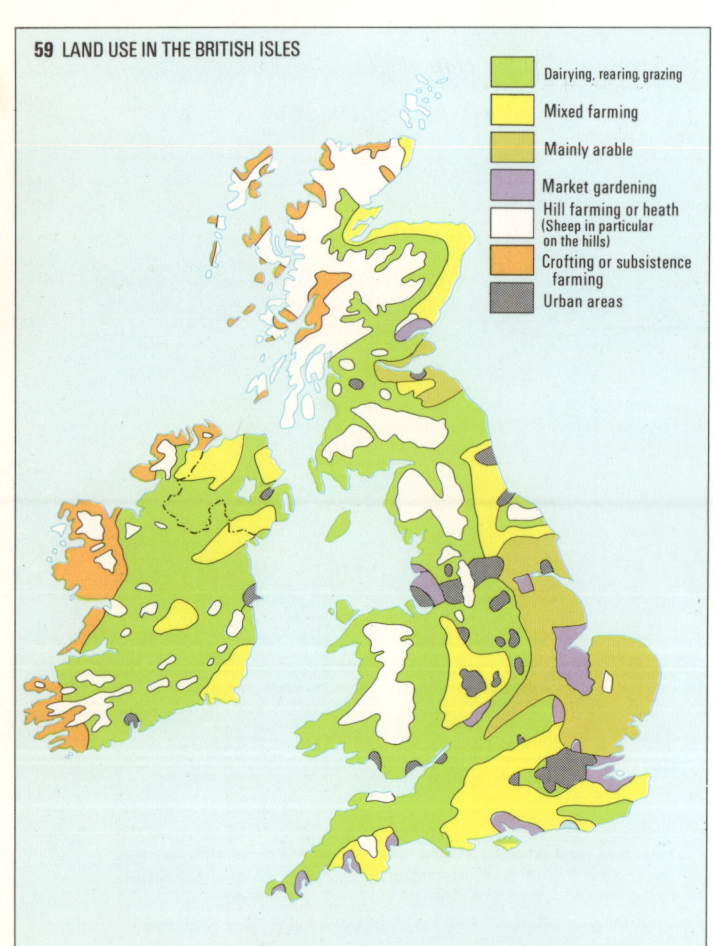

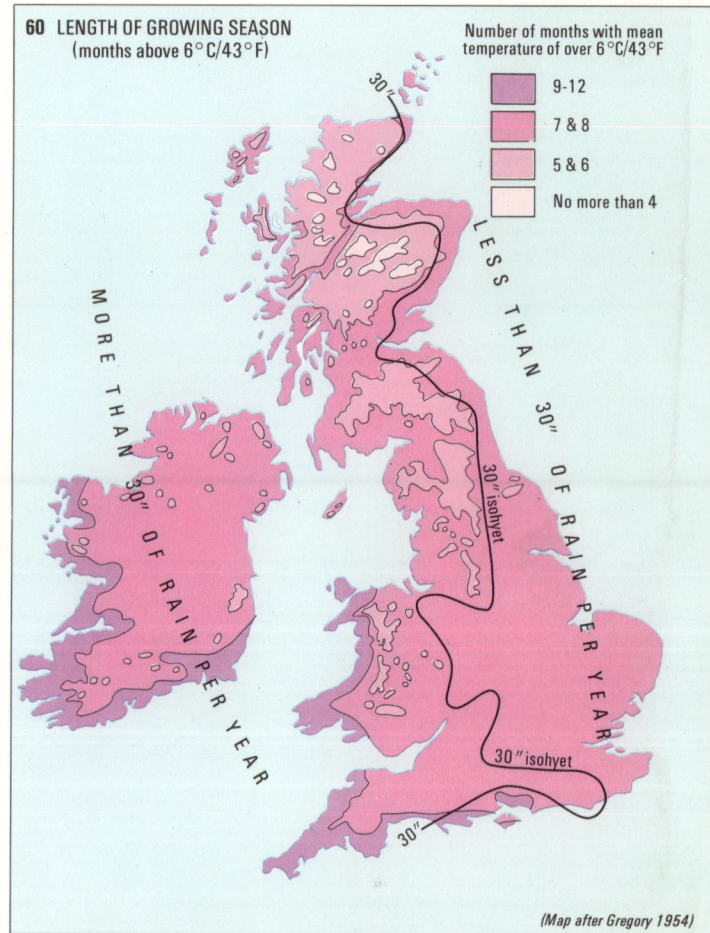

Figs. 59 and 60 These maps should be examined together.

59/60a What relationship do you notice between
  (i) the 'hill farming' or 'heath' areas, rainfall, and the length of the growing season?
  (ii) the main 'arable farming' areas, rainfall, and the length of the growing season?

59/60b Write a note on the meaning of 'crofting' and 'subsistence farming', and then describe the location of these in the British Isles.

59/60c How would you describe, very broadly, the farming of
  (i) the wetter west of Britain,
  (ii) the drier east?

59/60d Describe and try to account for the location of the areas which have the longest growing season.

BRITISH ISLES   Main crop distribution   11

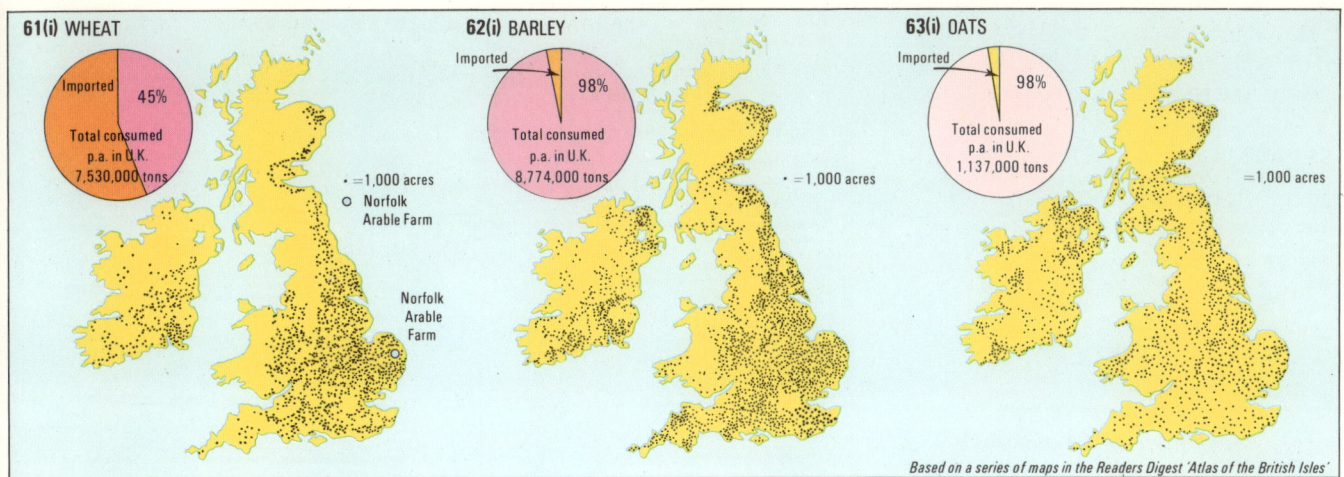

Figs. 61, 62 and 63   Cereal growing is encouraged by high summer temperatures, rainfall of about 25 in. per year or less (mainly in the spring), rich mixed soils, and flat land for machinery.

61a   Describe the location of the main wheat areas.
61b   Broadly, which areas have little wheat, and why?
62a   How does the distribution of barley (which is a little more tolerant of wetter and cooler climatic conditions than wheat) differ from that of wheat?
63a   Oats are more tolerant of wetter conditions than either wheat or barley. How is this shown by Fig. 63?
63b   What are the climatic conditions (summer temperatures, length of growing season, sunshine hours, annual rainfall) which encourage cereal cultivation in eastern United Kingdom? (Give actual figures.)
63c   How do conditions of relief and soil encourage cereal growing in these eastern areas? (Examine Figs. 14 and 22.)
63d   In what ways are the low temperatures of eastern UK in winter a help to the arable farmers there?
63e   Place the three cereals in order of importance according to total amount consumed in the UK.
63f   Which cereal is imported in quantity? Name two countries from which it comes.

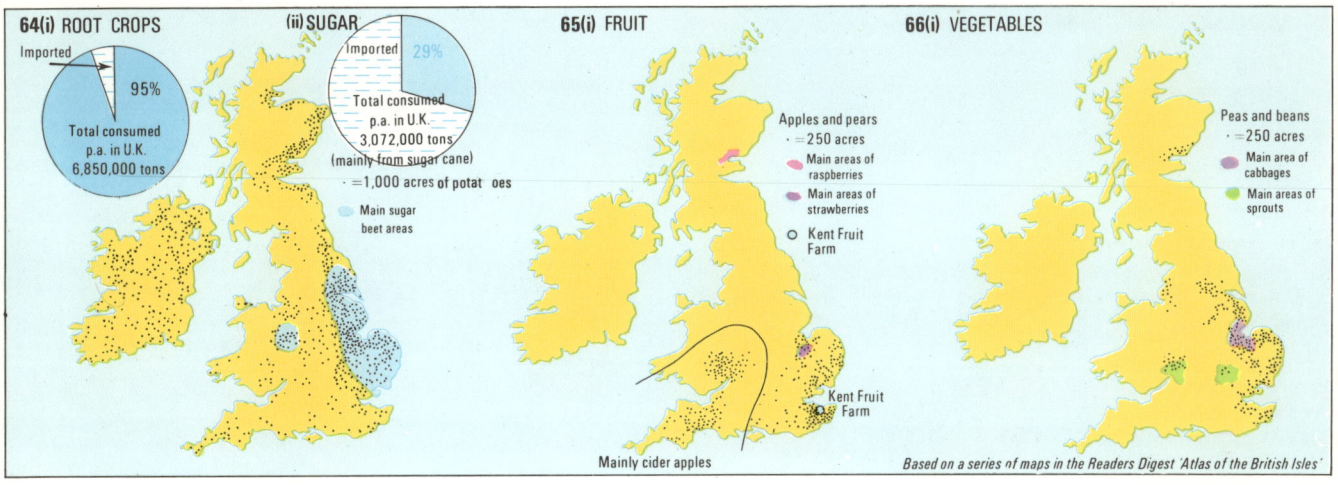

Fig. 64   Potatoes
64a   Describe, as precisely as you can, the main areas for potatoes (i) in England, (ii) in Scotland.
64b   The dry east and the deep black soils of the Fens favour potatoes and sugar beet there. What advantages have the eastern lowlands of Scotland for potatoes?
64c   From which country does the UK import most sugar?

Fig. 65   Fruit
65a   Describe as precisely as you can the location of (i) raspberries in Scotland, (ii) strawberries in England, (iii) fruit in south-east England.
65b   Describe the location of the main cider apple areas.

Fig. 66   Green Vegetables
66a   Suggest a name for the main area of cabbage growing, and name the main areas for brussels sprouts.

Fig. 67   Sample farms
67a   Describe and try to account for the main differences and similarities between the Norfolk arable farm and the Kent fruit farm.

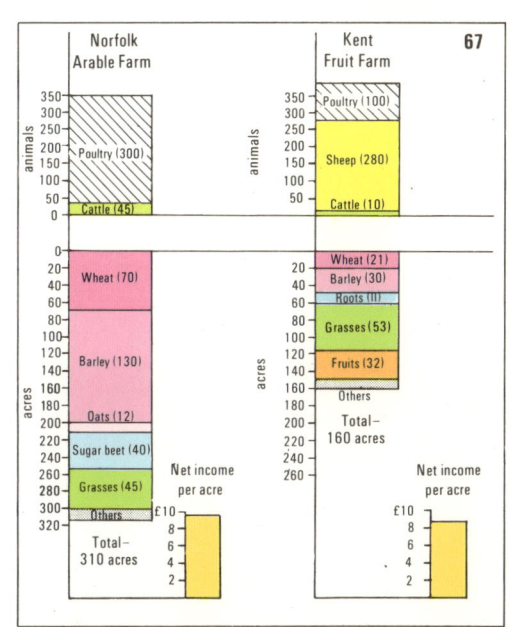

12 ENGLAND AND WALES Farming statistics to show general trends

A. FARM SIZE AND SOME MAIN CROPS

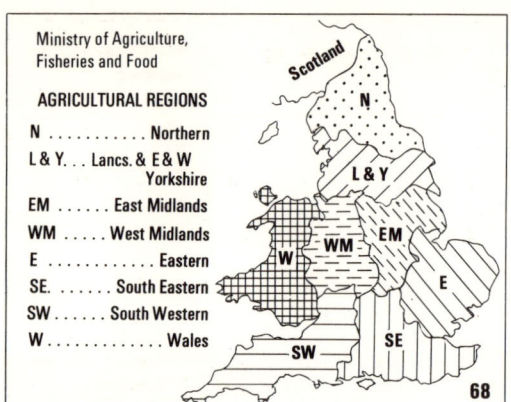

Fig. 68 The agricultural regions adopted by the Ministry of Agriculture, Fisheries and Food

68a Study your atlas map of England and Wales and then name the counties in each of the eight regions.

Fig. 69 Farms less than 100 acres

69a Examine the columns in Fig. 69 and then state the average percentage, for all England and Wales, of farms under 100 acres in area.

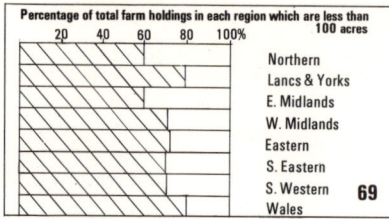
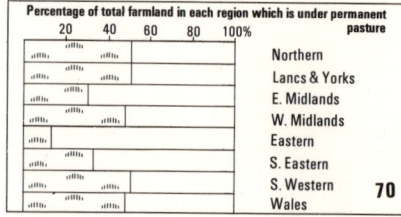
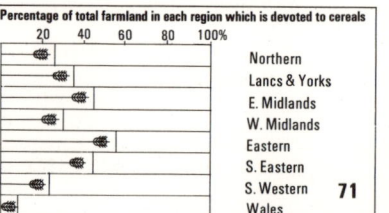
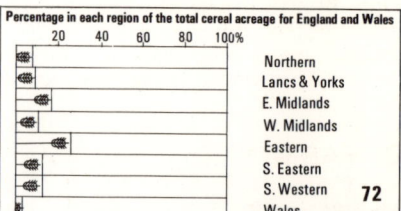

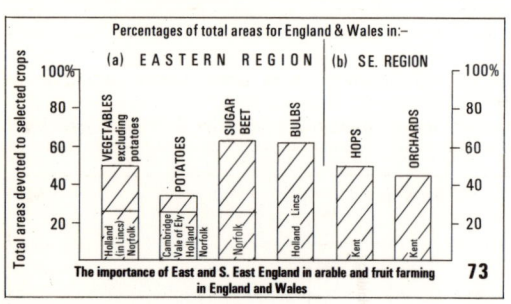

Fig. 73 Arable and fruits in two important regions

73a Describe the information given in Fig. 73 (e.g. which region has most sugar beet, potatoes, bulbs, hops, and orchards? Name areas within regions. What percentages of total areas devoted to the crops mentioned are found in the regions shown?)

B. LIVESTOCK

Fig. 74 Livestock in regions

74a What conditions of physical geography encourage sheep rearing (and/or discourage other agricultural activity) in Wales and the northern regions?

Fig. 75 Dairy herds in regions

75a What advantages does the South-West region enjoy which have encouraged its large dairying industry? (Think of grass growth, temperatures in winter, rainfall, transport, demand.)

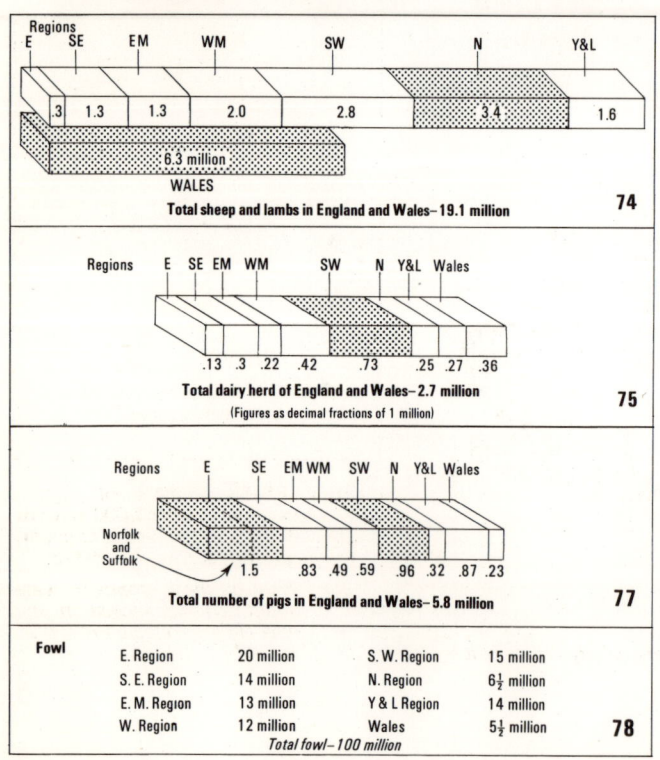

Fig. 70 The percentage of farmland in each region which is under permanent pasture

70a List (i) the five regions which have about 50% (or over) under permanent pasture (put them in order with the largest percentages first),
(ii) the three regions having less than 40% under permanent pasture (put them in order with the least percentage first).

70b Suggest two reasons for the low percentages under permanent pasture in the eastern regions, and for the high percentages in the western regions.

Fig. 71 Percentage of farmland in each region which is devoted to cereals

71a Which regions have over 40% of their total farmland devoted to cereals?

71b Are the regions with over 40% under cereals also those which have the highest percentages of total cereal acreage for England and Wales (see Fig. 72)?

Fig. 72 Percentage in each region of the total cereal acreage for England and Wales

72a Which two regions have the greatest percentages of the total cereal acreage in England and Wales?

72b Suggest two conditions of physical geography which have discouraged cereal cultivation in the two regions which have the lowest percentages of the total cereal land of England and Wales.

72c Give three advantages for cereal cultivation in the two leading cereal regions. (Think of temperatures, rainfall, soils, relief.)

Fig. 76 Beef cattle in areas

76a The three main regions for beef cattle are named in Fig. 76. How much larger is the dairy herd in each of these three regions?

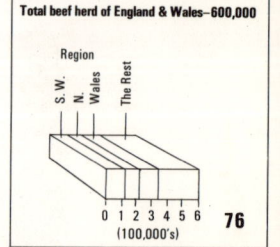

Fig. 77 Pigs in regions

77a Suggest a reason for the large number of pigs in (i) the South-West region, (ii) the East region. (Also look at Fig. 89 for production of bacon and ham.)

Fig. 78 Fowl by regions

78a Illustrate by means of bar graphs the figures for poultry given in Fig. 78.

78b Poultry-breeding today uses highly concentrated methods. What are these?

GREAT BRITAIN Farming statistics to show general trends

C. YIELDS

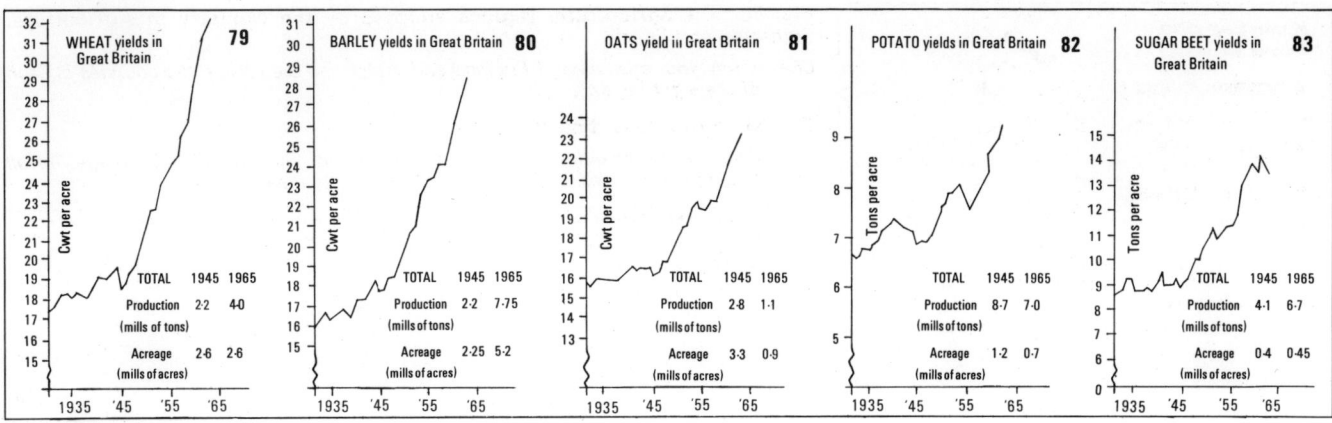

Figs. 79–83 The general trend is clearly for a very considerable increase in the yield of all crops over this 30-year period.

79–83a Which crops show an increase in *total output even though their crop acreages have increased only a very little or not at all*?

79–83b Which is the only crop which has had a marked increase in acreage?

79–83c Which crop has suffered a very considerable reduction in acreage?

79–83d Has the acreage of barley increased at a greater rate than total output?

D. FARM SIZE

Figs. 84 and 85

84/85a These graphs all indicate the same trend. What is it? The trend has continued in the 1970s.

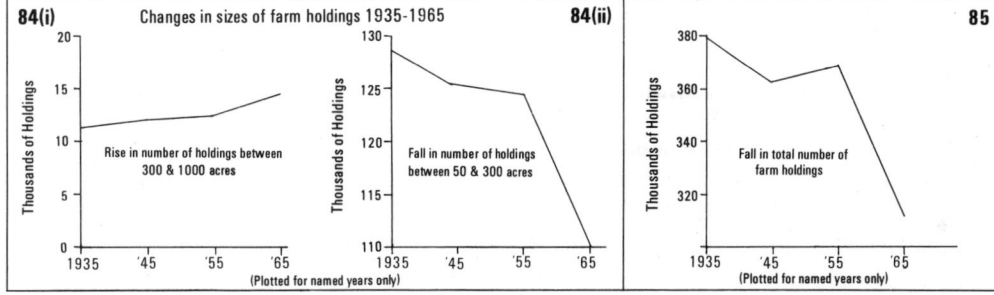

E. FARM WORKERS AND MACHINES

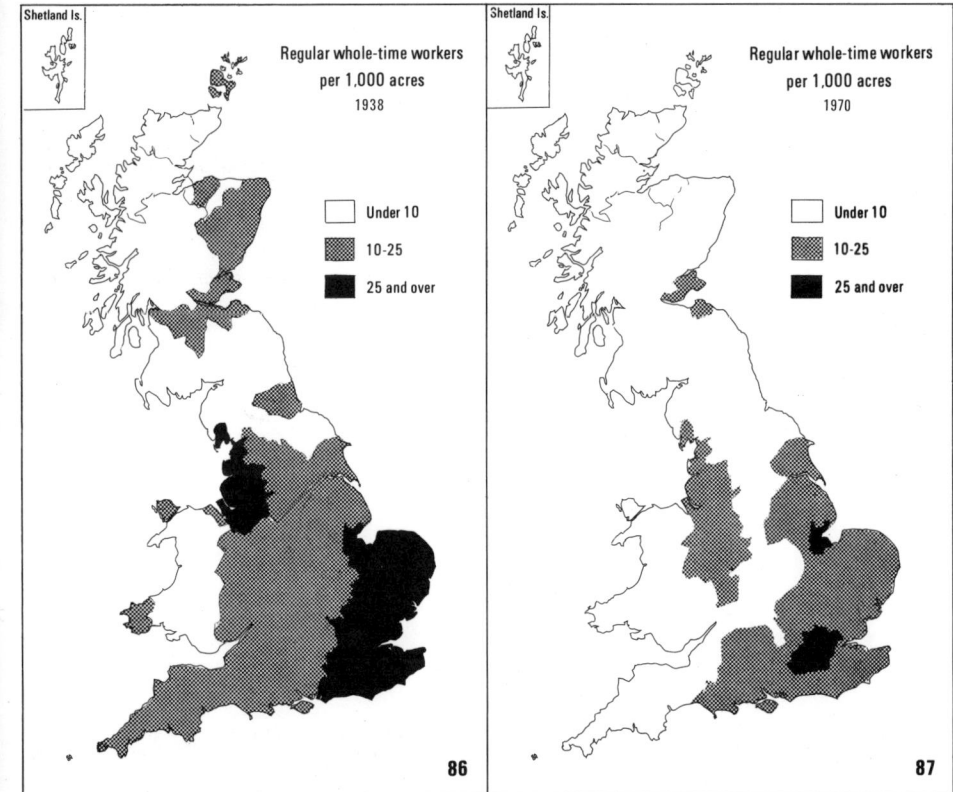

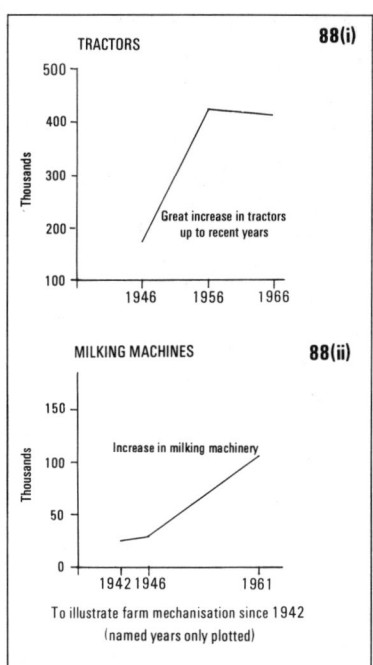

Figs. 86–88 Number of regular wholetime workers per 1,000 acres of agricultural land (including common rough grazing) and mechanisation.

88a What do these graphs indicate about farm mechanisation, and how are they related to Figs. 86 and 87?

Figs. 86 and 87

86/87a A comparison of these two figures brings out one main fact very clearly. What is it?

14 BRITISH ISLES Distribution of Livestock

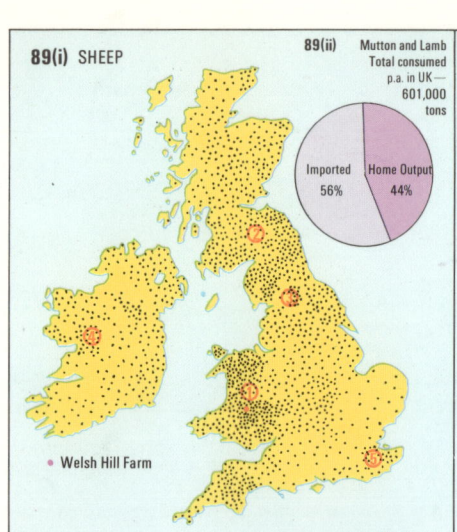

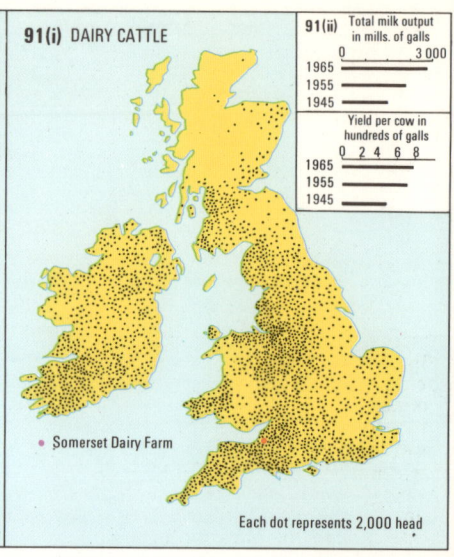

Figs. 89, 90 and 91 Sheep and cattle distribution

89/90a Name two countries from which the UK imports (i) mutton and lamb, (ii) beef and veal.

89/90b Name the main sheep areas shown on Fig. 89 and numbered 1–5.

90/91a What climatic conditions encourage cattle rearing in the western parts of the UK?

90/91b Which areas are important for both beef and dairy cattle?

90/91c Use your atlas to find the following: The Golden Vale of Limerick (S.W. Ireland), the Vales of Crediton and Taunton (S.W. England), the Plain of Cheshire and South Lancashire (north central England), the western side of the Central Lowlands of Scotland. What is the main livestock in each?

90/91d Describe, as clearly as you can, the location of the main beef cattle areas in (i) Scotland, (ii) Wales, (iii) England, (iv) North and South Ireland.

90/91e Which area is least important for both beef and dairy cattle? Suggest reasons for this.

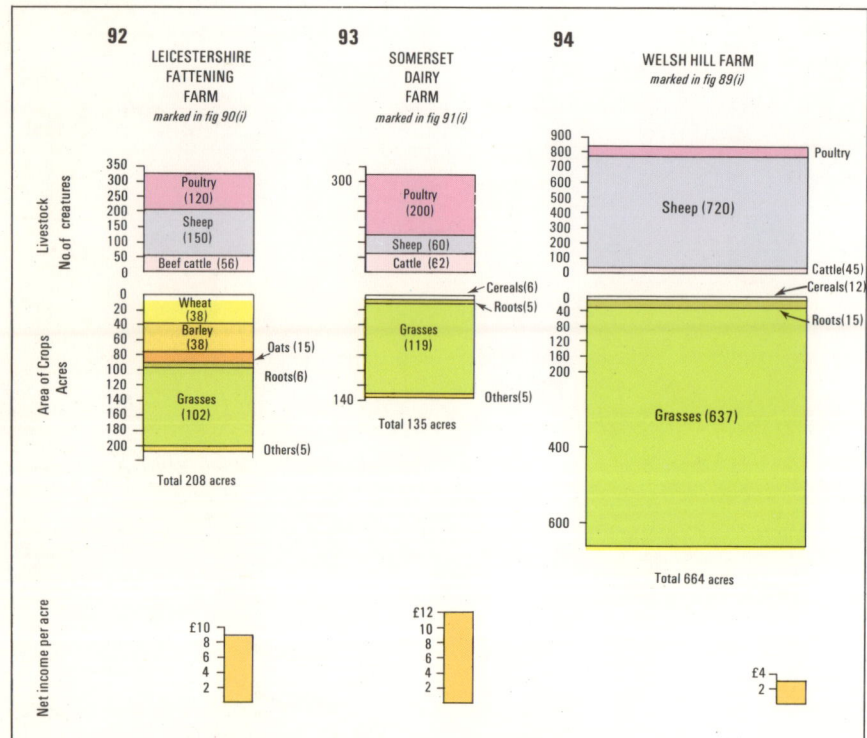
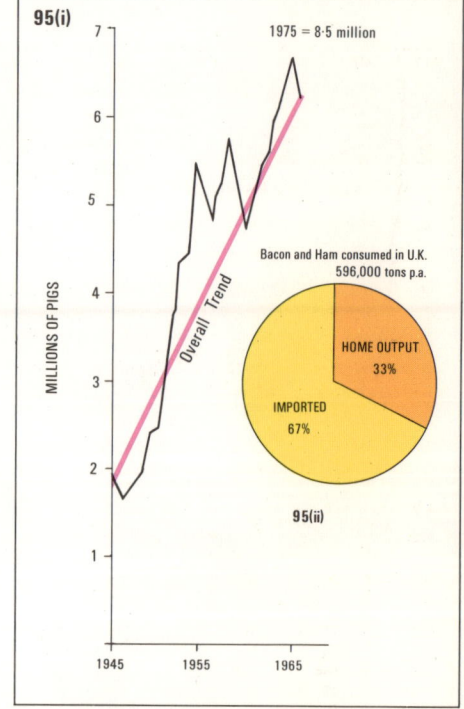

Fig. 95 Pig population trends

95a From which countries in particular does the UK import bacon and ham?

95b About how many thousand tons of bacon and ham are imported per year?

95c (i) What was the total number of pigs in the UK in 1965 and in 1975?
(ii) By what percentage has the pig population increased in the UK since 1945?

Figs. 92, 93 and 94 Farm incomes

92/94a Describe the conditions which favour the main farming activity on each of the three farms.

92/94b About how many cattle are there per acre of grass on (i) the Leicestershire Beef Farm, (ii) the Somerset Dairy Farm?

92/94c How many sheep per acre are there on the Welsh Hill Farm?

92/94d What is the total income for each farm?

92/94e By how much do the net incomes per acre differ between each of the three farms?

GREAT BRITAIN   The fishing industry

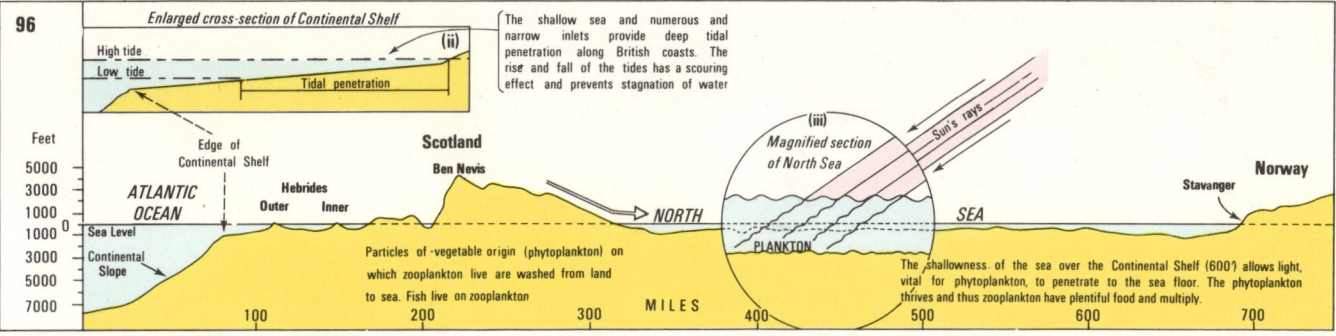

Fig. 96 There are three main types of fish: demersal (deep water), pelagic (shallow water), and shellfish. The deep-water industry is concentrated in Hull, Grimsby and Fleetwood, which house most of the distant-water ships and freezer trawlers. These trawlers have led to an increase in deep-water fishing in the waters off Newfoundland, Labrador and Greenland.

Near- and middle-water vessels, and herring drifters are concentrated in the Scottish fishing ports, and in a number of smaller ports around the coast of Britain.

Fish dispatch from these ports to inland centres (e.g. Billingsgate in London) is by express train or lorry, and quick-frozen products, by such firms as Findus and Birds Eye, are very important to the industry today.

The facts and figures relating to the British fishing industry are being altered all the time after the recent Cod War with Iceland, and by the demands of various countries and official pressure-groups for changes in present fishing rights. The pattern for the future is not yet clear.

96a Draw a sketch to show tidal penetration along a steep coast.

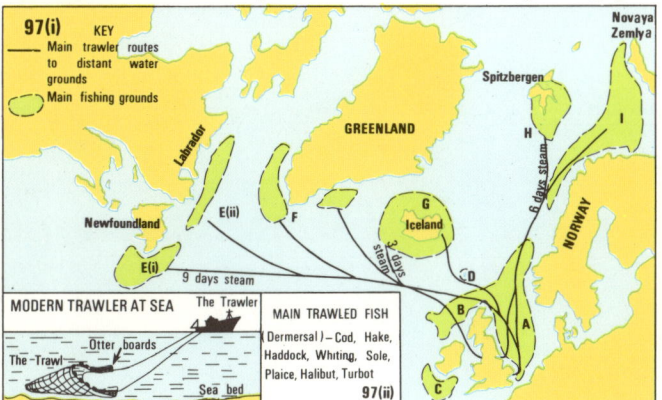

Fig. 97 Fishing grounds
97a Name the fishing grounds lettered A–I.

Fig. 98 Fishing ports
98a Name the fishing ports numbered 1–7.
98b Describe the main differences between dermersal fishing and pelagic fishing.

Fig. 99 Value of catch by regions
99a What is the value of the fish landed at Scottish ports from (i) the North Sea, (ii) West Scotland?

Fig. 100 Value of catch by ports
100a What is the total value of cod landed by British vessels?
100b Place the three leading ports of (i) Scotland, (ii) England and Wales in order according to value of fish landed.
100c What advantages do you think are enjoyed by Grimsby, Hull and Aberdeen?

Fig. 101 Value of catch by type of fish
101a What is the total value of:
  (i) all dermersal fish landed.
  (ii) all canned salmon imported into Britain.

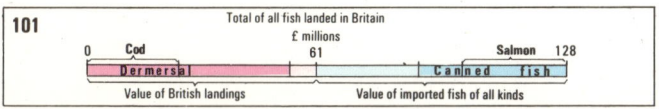

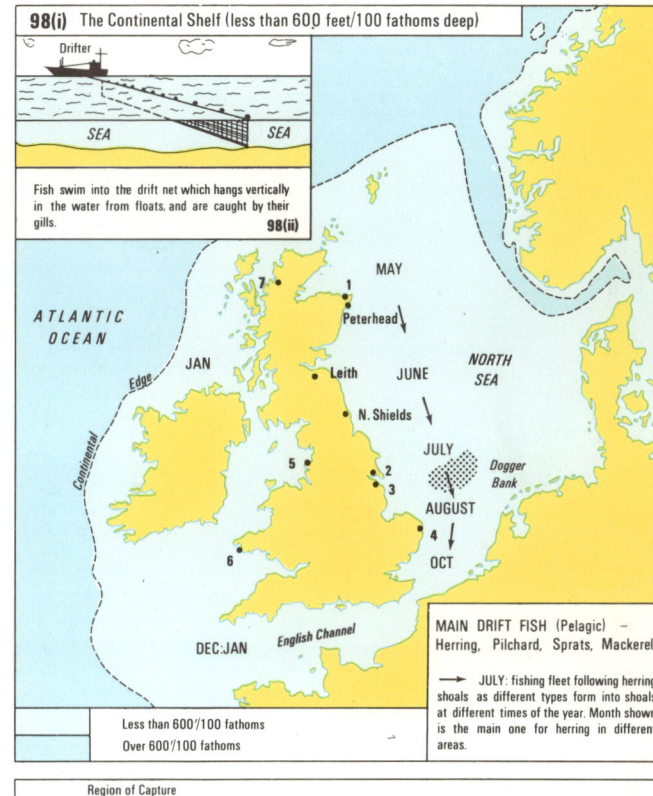

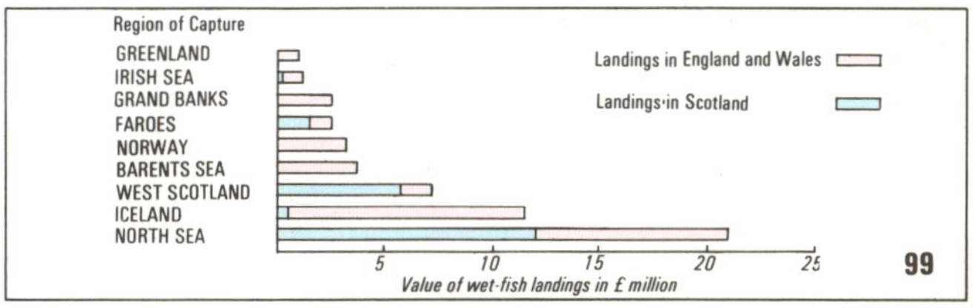

Fig. 102 Value of imported fish
102a Which was the most important fish imported into Britain?
102b Find out the main source of this fish for Britain.

The total value of all fish and fish-preparation exported from Britain in a recent year was £9·5 million — most of it frozen, chilled, or prepared in some way.

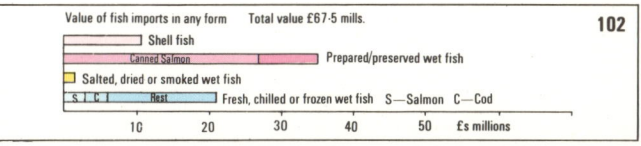

16 GREAT BRITAIN Coal: statistics showing general trends over a 20-year period

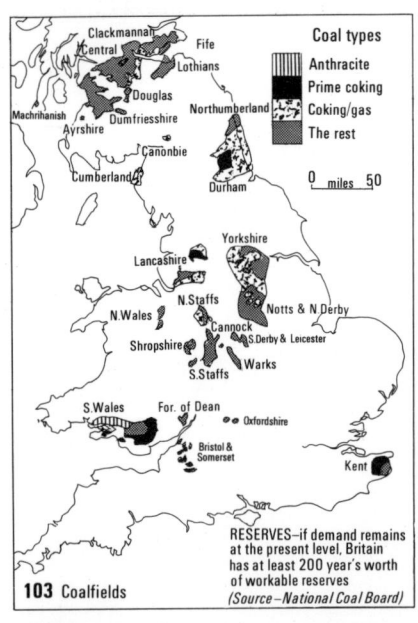

103 Coalfields

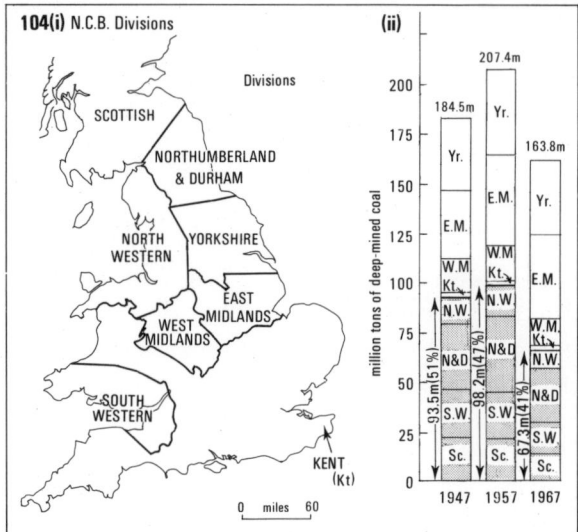

Fig. 103 Coalfields
103a First note the names and locations of the coalfields of Scotland, then of England, then of Wales.
103b Which coalfields have the largest areas of (i) anthracite, (ii) prime coking coal, (iii) coking/gas coal?
103c Describe the location of the Scottish coalfields.

Fig. 104 National Coal Board Divisions and Areas
104a Note the National Coal Board coalfield Areas.
104b What percentage of total output came from the Yorkshire and East and West Midland Areas in 1947 and in 1967?
104c List the three most important Areas today and, using the scale line, estimate the output from each.

The N.C.B. policy is to concentrate output on the most productive Areas.

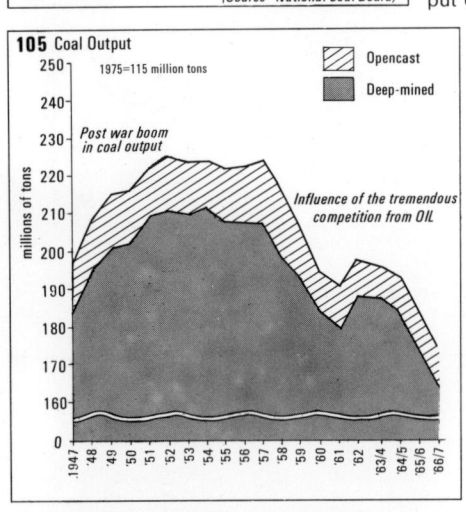

Fig. 105 Coal output
105a Why was there a 'coal boom' (i.e. a period when demand was in excess of immediate supply at the old prices) just after the 1939/45 war?
105b Describe the influence of cheap oil imports and use on coal supplies.
105c What was the total coal output in 1947, 1957, 1967 and 1975?

Fig. 106 Number of collieries
106a How many collieries were there in 1947, 1957 and 1967?
106b List in three vertical columns the number of collieries for each Area in the years 1947 and 1967 and next to 1967 write the number by which the total of collieries had fallen since 1947.
106c Which Areas have had over 100 collieries closed since 1947?

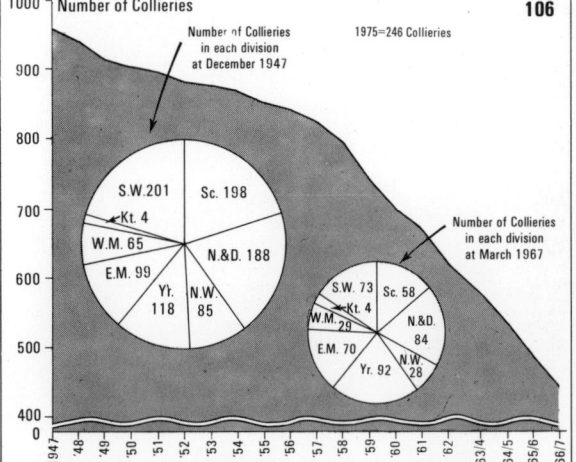

Fig. 109 Mechanisation in the coal industry
109a What is the main fact brought out by Fig. 109?
109b What percentage of total output was mechanised in 1947, 1967 and 1975?

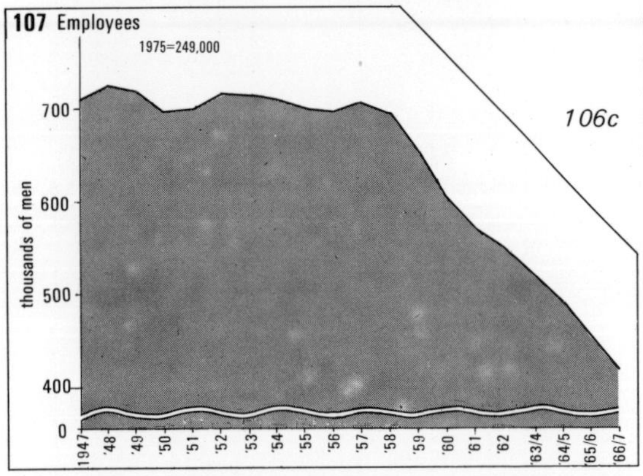

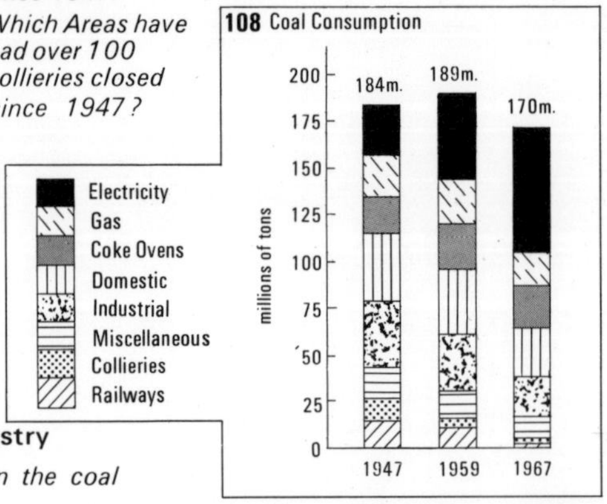

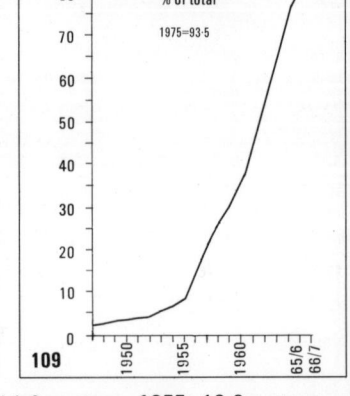

Fig. 107 Number of men employed in the coal industry
107a When did the big decline in numbers employed in the coal industry begin?
107b Suggest reasons for this decline.
107c How many were employed in 1957 and in 1975?

Fig. 108 Coal consumption
108a Describe, and try to account for, the changes in consumption by the following coal-users between 1947 and 1967: electricity, domestic, collieries, railways.

110:— **1947**: 5·3 mill. tons   **1951**: 14·9 m. tons   **1955**: 13·9 m. tons
      1959: 9·7 m. tons   **1963**: 8·2 m. tons   **1967**: 2·5 m. tons

Fig. 110 Total coal exports
110a Graph these figures and add the figure 2·0 m.tons for 1975.
110b Describe the changes in export, and try to explain the general trend.

BRITISH ISLES Electricity and gas 17

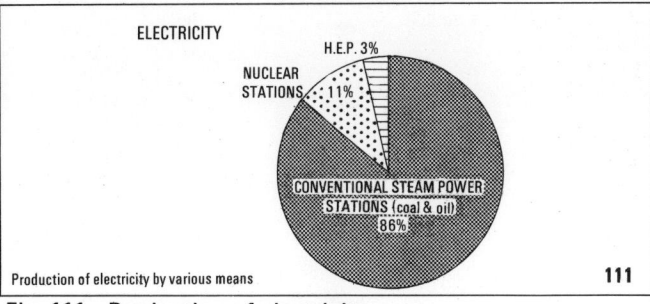

Fig. 111 Production of electricity
111a Electricity is made in several ways: place them in order of importance in Britain.
111b What is meant by Conventional Steam Power?

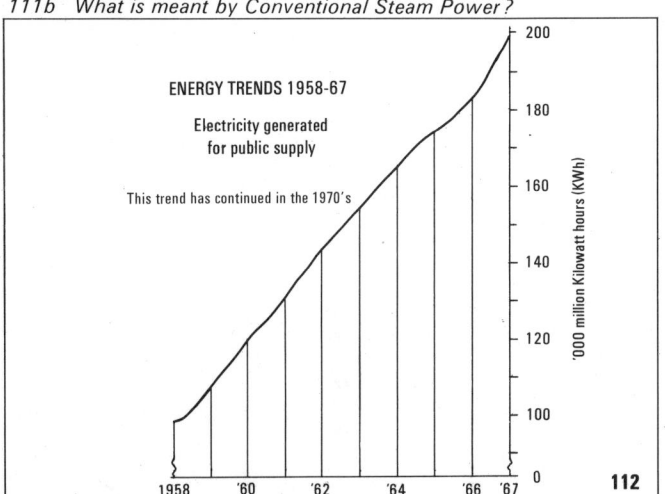

Fig. 112 Energy production
112a How much greater was the supply of electricity in 1967 than in (i) 1965 and (ii) 1963?

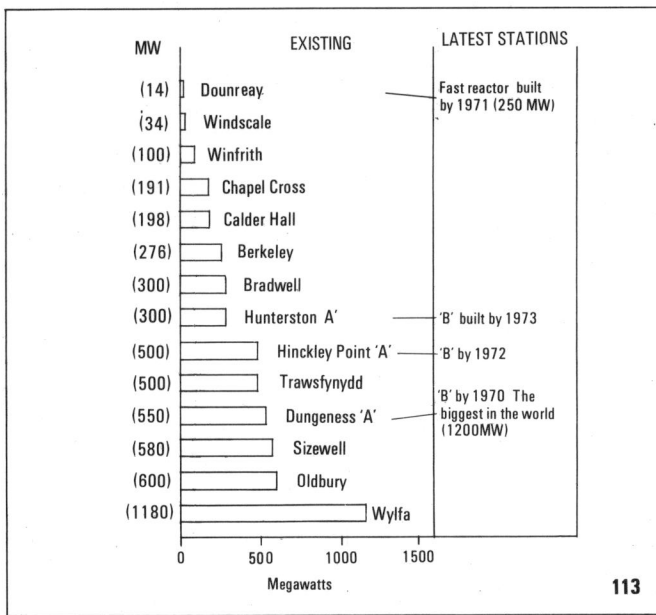

Fig. 113 Nuclear power stations
113a The supply of electricity from nuclear power stations has increased greatly. Which stations supply more than 300 megawatts?
113b Which stations named on Fig. 113 have not been marked on Fig. 114? Make sure you can place them.

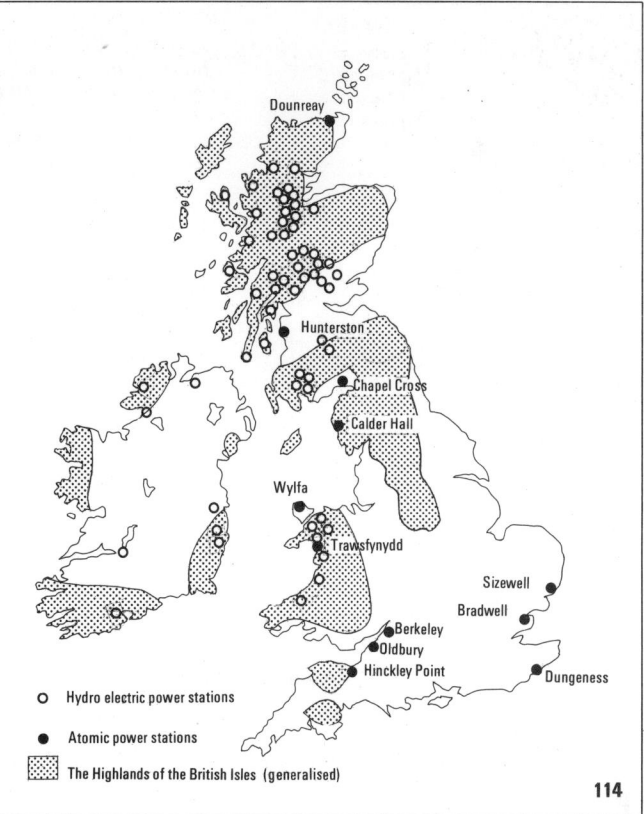

Fig. 114 Hydro-electricity
114a What do you notice about the general location of (i) hydro-electric power stations (ii) the atomic power stations?
114b Suggest the advantages of the glaciated highland areas of the British Isles for H.E.P. production (think of relief and its special features, and of climate).

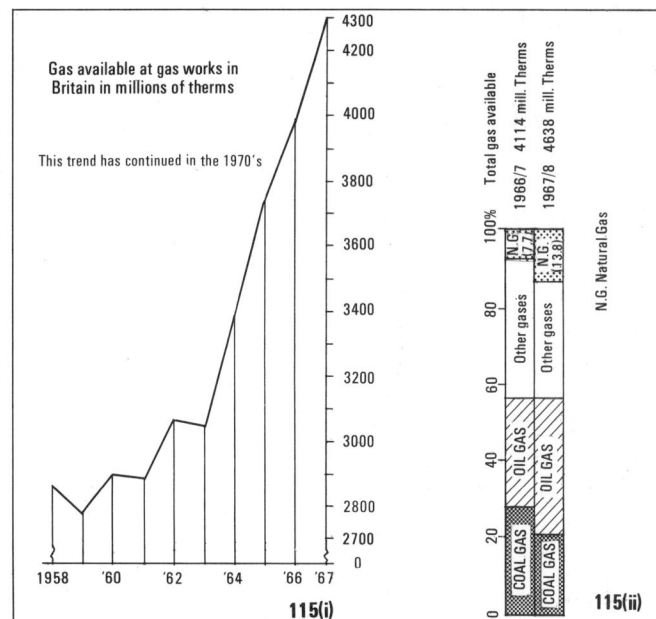

Fig. 115 Gas
115a How much greater was the total of gas available in Britain in 1967 than in (i) 1965 and (ii) 1963?
115b What does Fig. 115 (ii) show about the relative importance of coal-gas, oil-gas and natural gas in 1966/7 and in 1967/8?

BRITISH ISLES Gas and oil

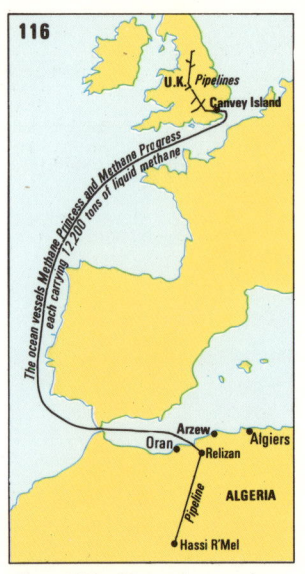

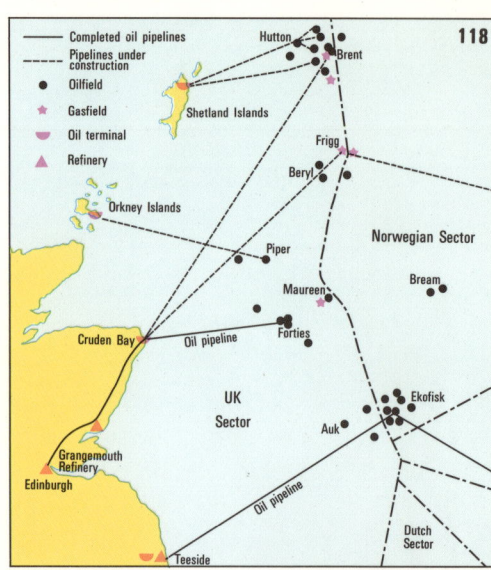

Fig. 116 Natural gas from Algeria accounts for about 10% of Britain's total gas supply.

116a Find out all you can about:
  (i) the two vessels 'Methane Princess' and 'Methane Progress',
  (ii) the Canvey Island methane terminal.

Fig. 117 The gas from the Sahara is pumped to eight Area Gas Boards along a grid network which cost about £10 million to build. The system has 200 miles of trunk pipeline and 150 miles of branch line.

117a Find the name of the nearest terminal installation to you in your named Gas Board Area.

117b Natural gas also comes from the North Sea. Study Fig. 117 and state
  (i) the names of the points at which the natural gas comes ashore,
  (ii) the names of the natural gas fields at present in production.

Fig. 118 North Sea Oil

118a List some of the British North Sea oil-fields.

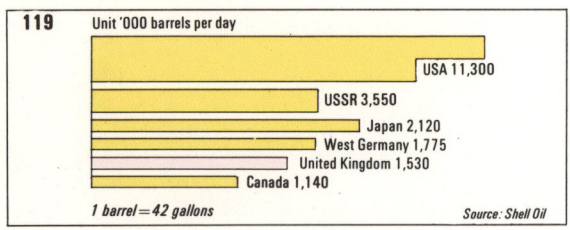
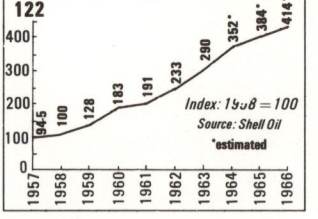
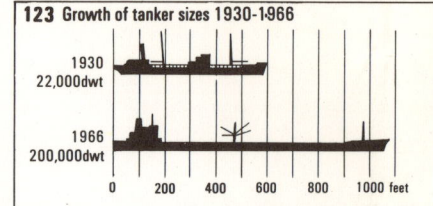

Fig. 119 Oil consumption, UK in a recent year

119a Taking the population of the United Kingdom to be approx. 54 million, work out the quantity (in gallons) of oil consumed in that year per head.

Fig. 122 Organic petrochemicals production

122a What is meant by the term 'organic petro-chemicals'?

Fig. 123 Tankers

123a Describe what is shown by Fig. 123.

123b What is the tonnage of today's largest tanker?

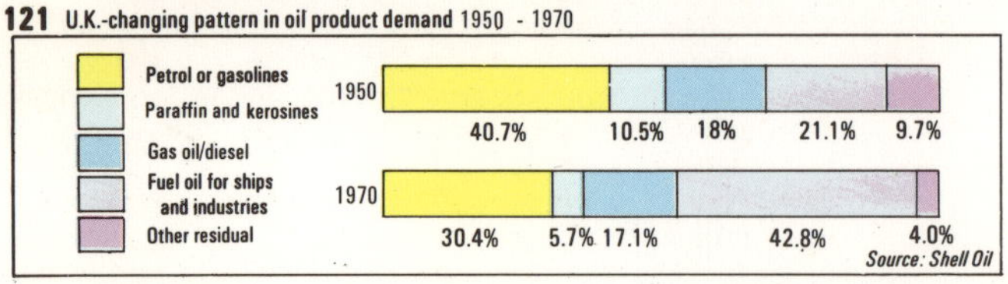

Fig. 124 Oil refineries

124a Describe the location of the oil refineries. Which ones can take the largest oil tankers?

Fig. 125 Oil consumption

125a What conclusions might be drawn from the figure?

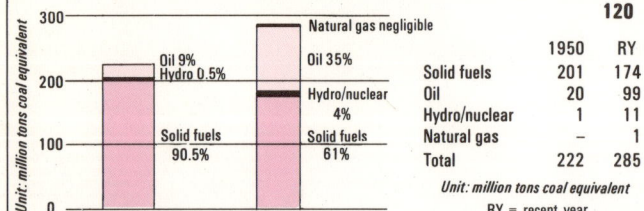

Fig. 120 Prime energy scene

120a Describe the changes which took place between the years illustrated here in the relative importance of the following solid fuels, oil, hydro-plus-nuclear energy.

Fig. 121 Oil product and demand

121a Total oil refined in Britain increased from about 20 million tons in 1950 to about 70 million tons in 1970. The percentages of the products of refining have not remained constant either. Suggest why a larger percentage of total refined product is made into fuel oil today and a smaller percentage into paraffin. Was the total amount of each made in 1970 smaller or larger than in 1950?

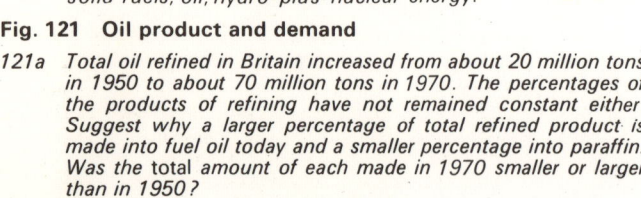
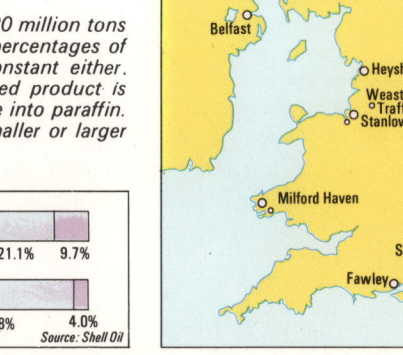
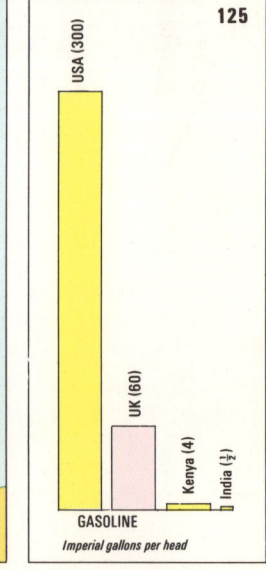

GREAT BRITAIN   Iron ore and steel output

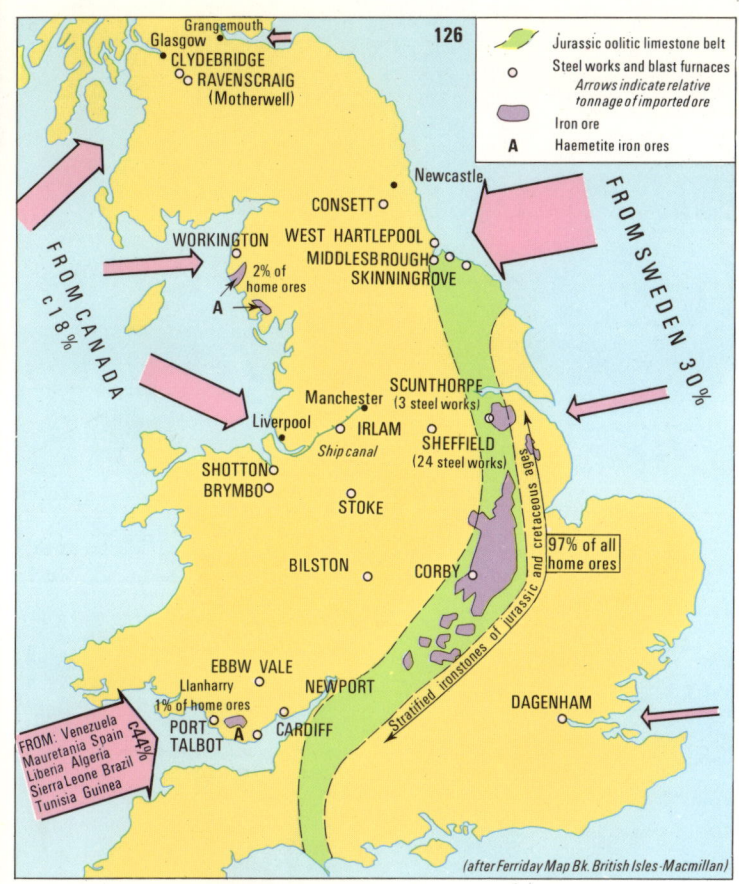

Fig. 126 Sources of iron ore, and conversion centres

Fig. 126 shows (i) the main areas of iron ore mining in Britain,
(ii) the main sources of imported ores,
(iii) the large centres of blast furnaces and steelworks.

126a Examine each of these groups on the map and note their names and locations.

126b Find out about (i) Abraham Darby of Coalbrookdale,
(ii) the Thomas-Gilchrist process (1879).
Explain the importance of each in the history of iron and steel making in Britain.

126c Examine Figs. 103 and 126 together and show how they are related. (Remember that traditionally coking coal is used for smelting.)

126d What part does limestone play in the industry? Suggest possible sources of this mineral for the industry.

Fig. 127 Output of British iron ores

127a Describe the changing pattern of production of the three main iron ores.

127b Why is it now thought that some of the early-established steel-making towns are no longer well placed (e.g. Sheffield), but remain very important producers because the skills and capital equipment are still there ('geographical inertia')?

127c Note the varying degrees of purity of the iron ores, and suggest why it is cheaper to take the coal to the Jurassic ores than to take the ores to the coalfields.

127	Output of British iron ores in '000 tons			
Kind and District	%Fe purity	1913	1965	1967
HAEMETITE				
Cumberland	49%	1,361	261	250
Glamorgan	50%	44	147	149
JURASSIC ORES				
Lincs (especially Scunthorpe)	24%	2,641	7,596	7,131
Northants (esp. Corby), Leicester, Oxford, Rutland	26% average	3,915	7,410	5,208
Cleveland Hills	28%	6,040	NIL	NIL
BLACKBAND (coalfield) ORES	average 33%			
Staffs		891	NIL	NIL
Central Scotland		592	NIL	NIL
Lancashire		406	NIL	NIL
TOTALS		15,990	15,414	12,738

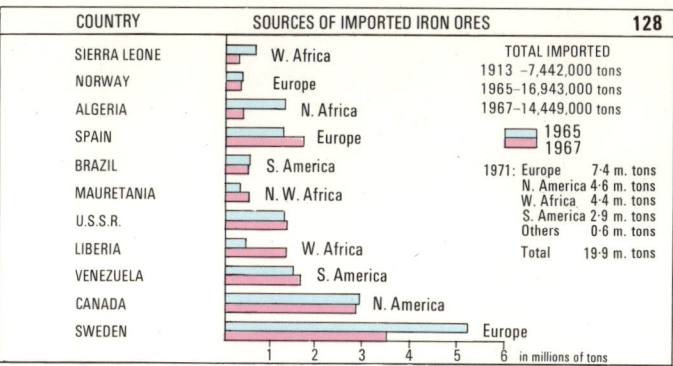

Fig. 128 Sources of imported ores

128a Have imported ores become more, or less, important since 1913?

128b What do you notice about the total iron ore used (home plus imported) in recent years?

128c Which supplying countries are becoming (i) more important, (ii) less important to Britain?

128d From which continent does Britain obtain most of her imported ores?

128e Why are some important new steelworks on or near the coast (e.g. Margam in South Wales)?

Fig. 129 Steel output trends

129a Which two districts lead in the production of steel?

129b By how many tons of output does the leading district exceed that of the second most important producer?

129c South Wales imports iron ore from abroad and pig-iron from home Districts numbers 1 and 7 (see map). Find out where the TIN comes from for the tinplate industry.

129d What is tinplate? Name some of its important uses.

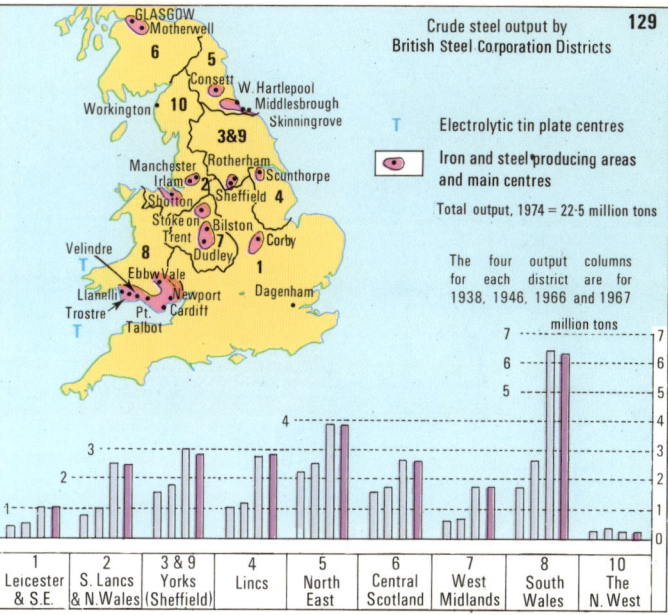

N.B. The B.S.C's Third Report on Organisation (1969) proposed that from March 1970 there should be a new structure, with the steel industry in four major Divisions:

(a) **Heavy Steel Division.** 85,000 employees: Headquarters at Glasgow and Teesside; output 10 million tons crude steel per year.

(b) **Special Steel Division.** 42,000 employees; H.Q. in Sheffield; output 3·8 million tons of high-priced steel per year.

(c) **Sheet Metal, Tinplate and Narrow Strip Division.** 69,000 employees; H.Q. in Cardiff; output 7·8 million tons per year.

(d) **Tube and Pipe Division.** 37,000 employees; H.Q. at Corby.

GREAT BRITAIN Textiles, ships, motor vehicles - general trends

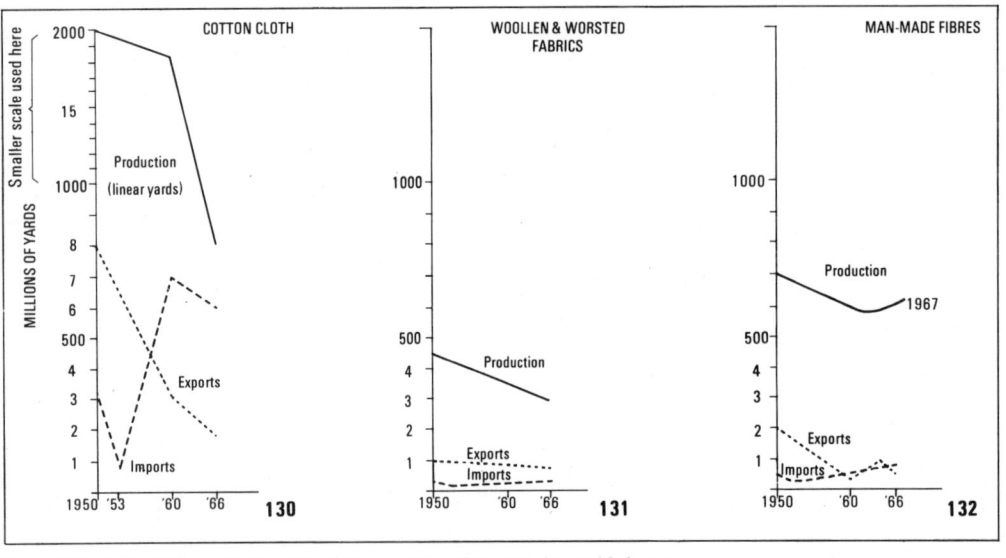
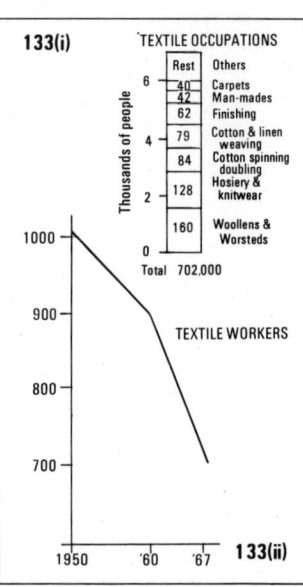

Figs. 130, 131, 132 and 133   Textiles: production, trade and labour

130a What changes are indicated in (i) production, (ii) imports and exports, of cotton cloth in Fig. 130?

130b Why have the changes taken place? (Think of 'cheap' labour in Japan, India and Hong Kong.)

131a Has the decline in production been as rapid in the woollen and worsted industry?

132a In 1950 man-made fibres accounted for 22% of total textile output. For what percentage did they account in 1967?

132b Nylon yarn production was 30 million lb. in 1953, 57 million lb. in 1959, 130 million lb. in 1963 and 400 million lb. in 1967. Illustrate these figures on a graph.

133a There has been a big decline in employment in textiles since 1950. Compare this with the decline in total textile output.

130–133a Find out the history of the three main textiles in UK (cotton, woollens, man-mades) and write an account of each (where developed, when, and why? reasons for changes?).

130–133b Write a paragraph on each of the following important names in the development of textiles: (i) Hargreaves, (ii) Crompton, (iii) Arkwright, (iv) Cartwright, (v) Eli Whitney, (vi) Courtauld.

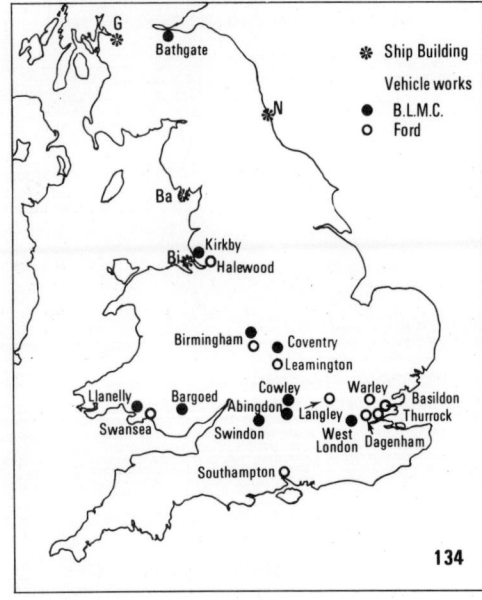
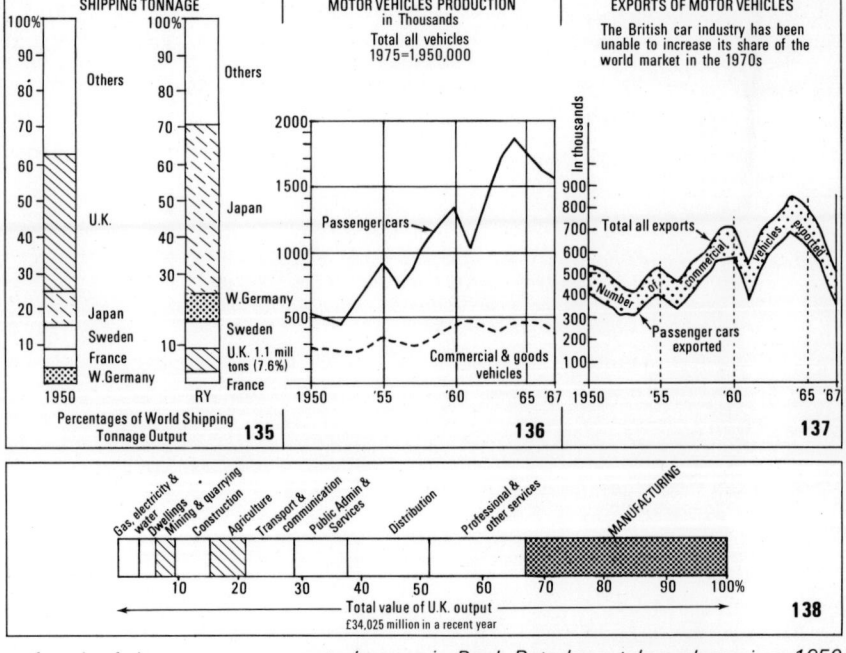

Figs. 134, 135, 136 and 137   Shipbuilding and motor vehicle manufacture

134a Find out and write a note on the advantages of each of the shipbuilding centres marked (name each).

134b Suggest names for the various groups of motor manufacturing centres.

134c Find out the centres at which Vauxhall and Rolls Royce carry on manufacturing.

135a Describe the main changes shown in Fig. 135 and try to account for the changing fortunes of the UK and Japan. (Think of relative wages, methods and capital equipment.)

136/137a The motor vehicle industry is sensitive to changes in Bank Rate* and the 'business climate'. Find out when important changes in Bank Rate have taken place since 1950. What effects have they had?

136/137b Although there have been fluctuations, what has been the general trend in vehicle output since 1950?

* Bank Rate is the interest rate charged by the Bank of England for loans. All other rates of interest usually follow changes in Bank Rate.

Fig. 138   Primary and manufacturing industries

138a Describe the relative importance of Primary (raw material) producers and Manufacturing shown in Fig. 138.

IRELAND Physical background

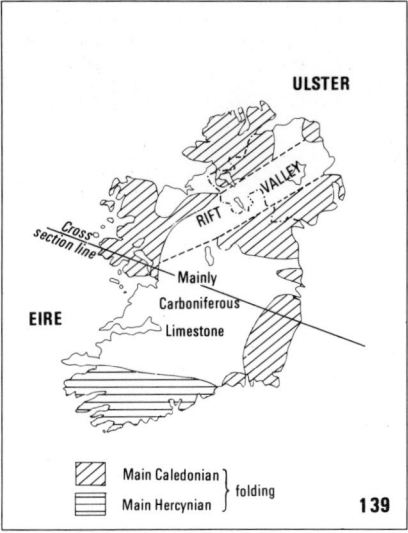
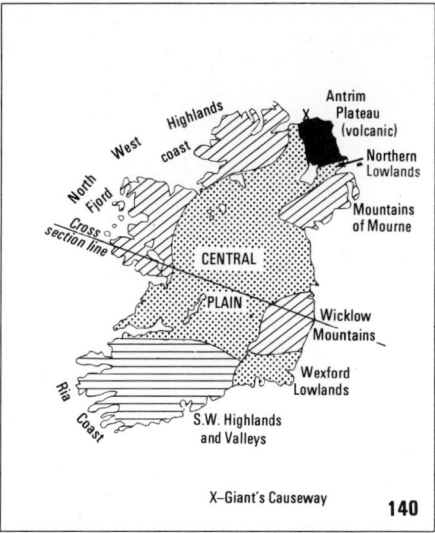
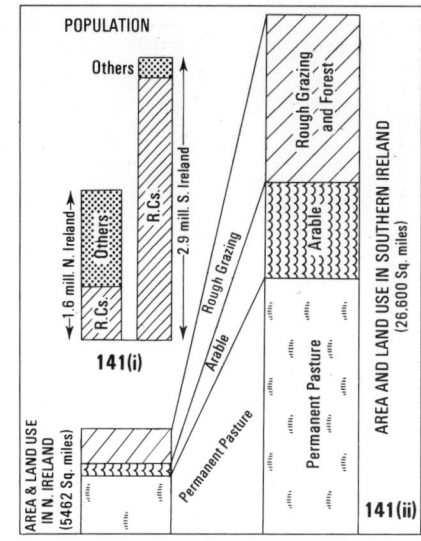

Figs. 139 and 140 Structural history and relief regions

139/140a Describe the relationship between Figs. 139 and 140.

Fig. 141 Population and land use

141a How much larger is the population of Southern Ireland than that of Northern Ireland?
141b Which part of Ireland has the more Roman Catholics?
141c Describe the characteristics of a *ria* coast and a *fjord* coast and mention particularly the ways in which they differ.
141d Give reasons for the very large percentages of land under permanent pasture in both North and South Ireland (think of climate).
141e Draw a sketch map of Ireland and mark on it the following features:
  (i) mountains: *Sperrin Mts., Donegal Mts., Mts. of Kerry and Cork,*
  (ii) rivers: *Shannon, Liffey, Erne, Bann, Foyle, Lagan,*
  (iii) loughs: *Lough Erne, Lough Neagh, Lough Derg,*
  (iv) bog: *The Bog of Allen.*
Use your atlas to help you.

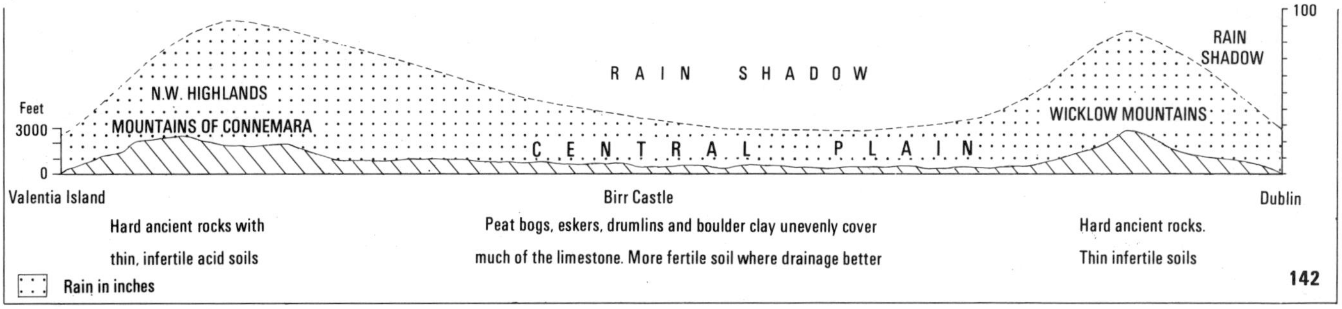

Figs. 142 and 143 Cross-section and climographs

142/143a The three graphs show some important differences between the west, the centre and the east of Ireland. State these differences and attempt to explain them. (Deal with this under temperature, rainfall and 'snow days'.)
142/143b Which of the three places has no months below average monthly temperature of 43°F/6°C? Give reasons for this.
142/143c Show how rainfall is related to relief as shown on the cross-section, Fig. 142.

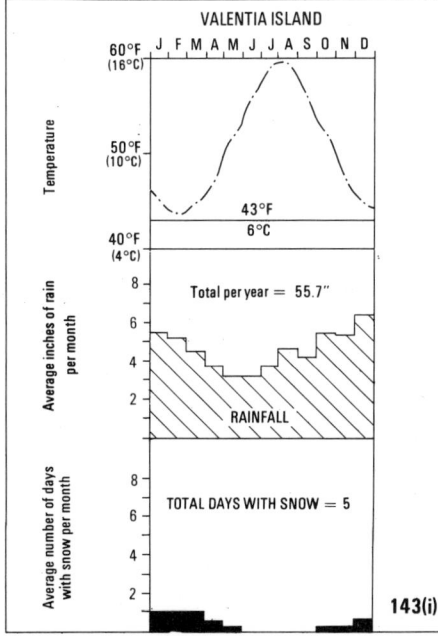
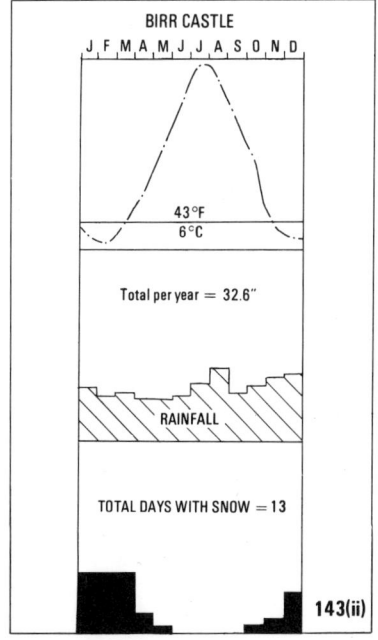
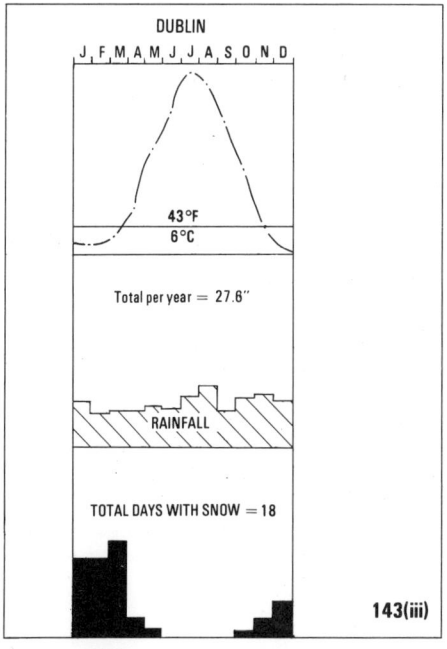

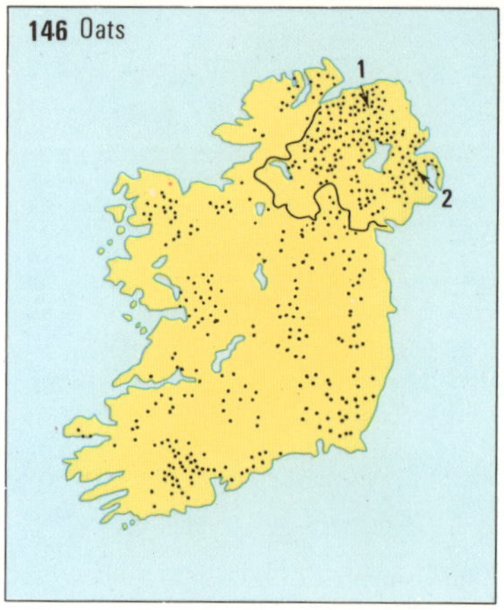

22 IRELAND Agriculture

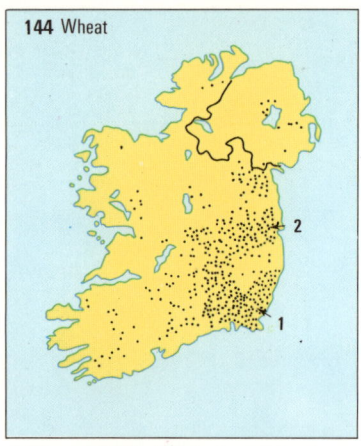

Figs. 144 and 145 Wheat and barley

144/145a Is there any relationship between Fig. 37 and Figs. 144/145?

144/145b Suggest names for the areas numbered 1 and 2 on Fig. 144 and numbered 1 on Fig. 145.

Fig. 146 Oats

146a Suggest names for the areas numbered 1 and 2.

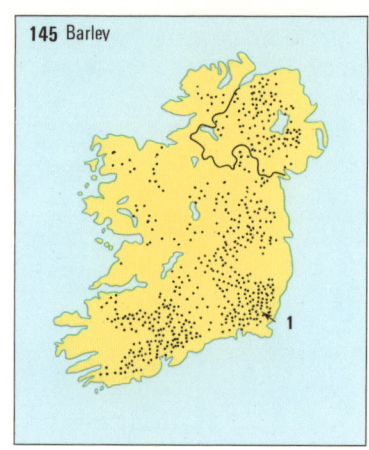

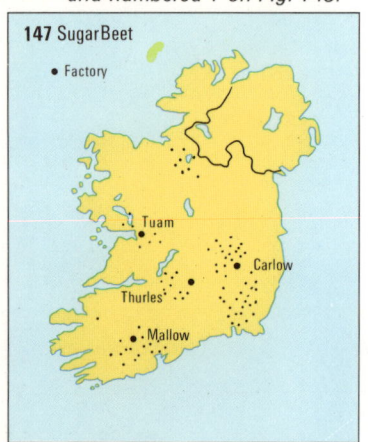

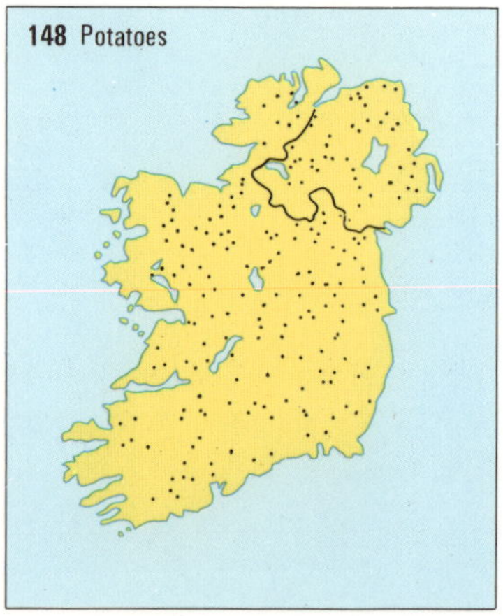

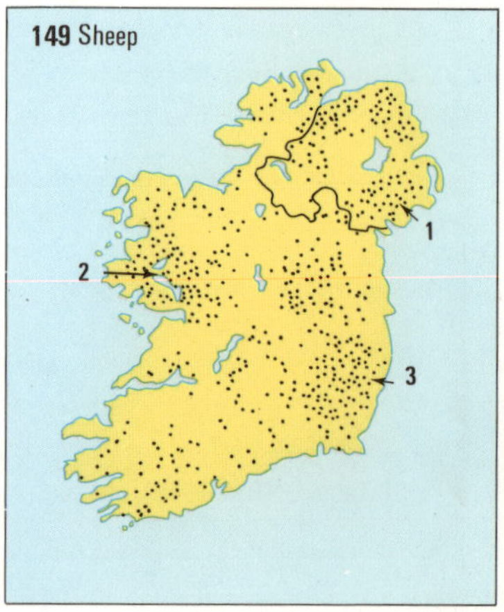

Fig. 147 Sugar beet

147a What are the uses of sugar beet?

Fig. 148 Potatoes

148a Describe the location of potatoes in Ireland.

148b What climatic factors encourage the growth of potatoes in Ireland?

148c What historical event of great importance took place in Ireland in 1846? What were its main results?

Fig. 149 Sheep

149a Suggest names for the areas numbered 1, 2 and 3. All three have sheep. What else do they have in common?

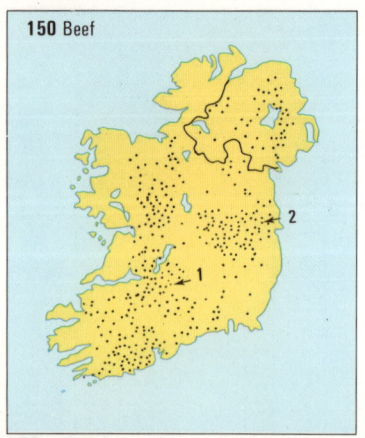

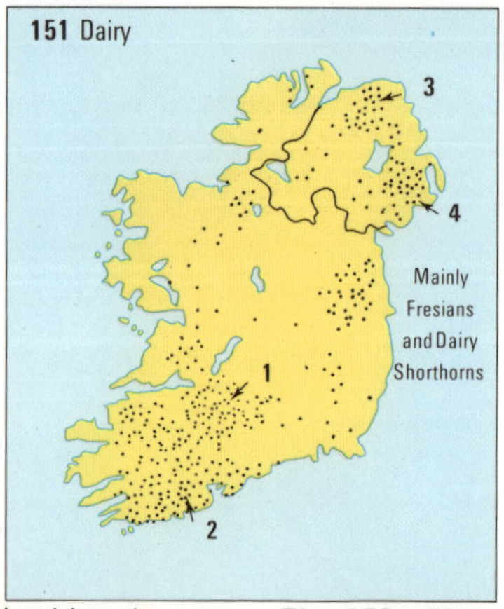

Fig. 150 Beef cattle are fairly widespread in Ireland but there are concentrations.

150a Where are they concentrated?

Fig. 151 Dairy cattle

151a Describe and try to account for the distribution of dairy cattle. (Think of temperatures and rainfall, and pasture and demand).

151b Find out what Friesian and Dairy Shorthorn cattle look like and describe each.

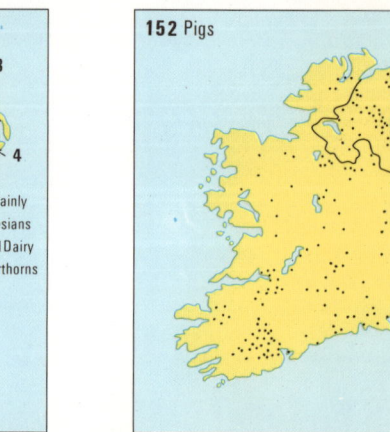

Fig. 152 Pigs

152a Is there any relationship between the distribution of pigs and of dairy cattle? If so, try to account for it. (What are pigs fed on?)

In the Western Highlands are 'cottage farmers' or 'crofters' who sell very little off their small farms, and produce oats and potatoes, cattle, pigs, sheep and poultry mainly for their own use. It is an area of high humidity and marshy peat-covered uplands, and many are migrating from it.

IRELAND Agricultural statistics to show trends

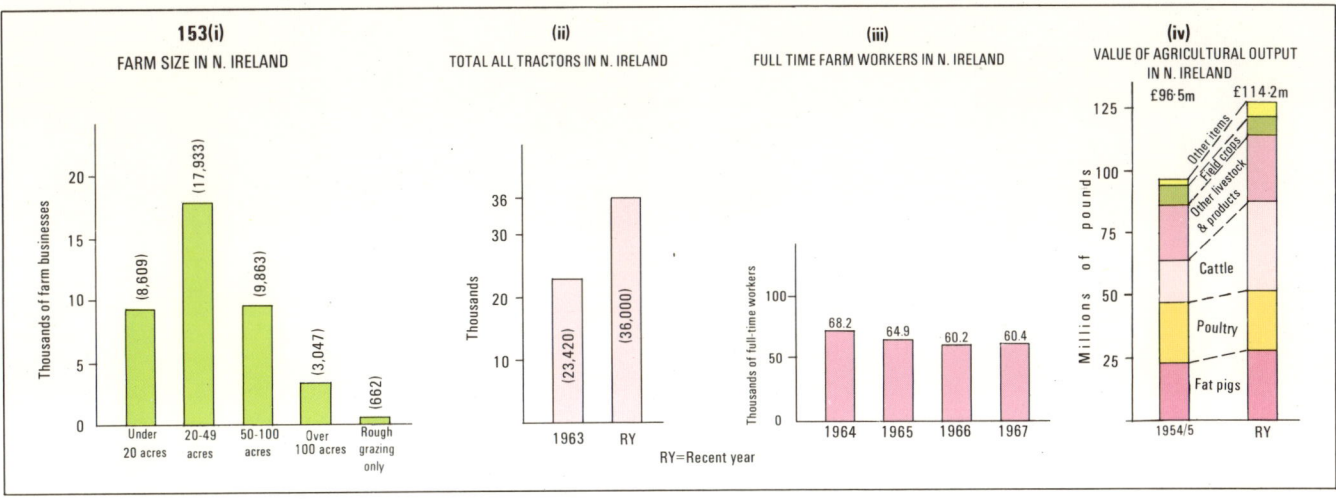

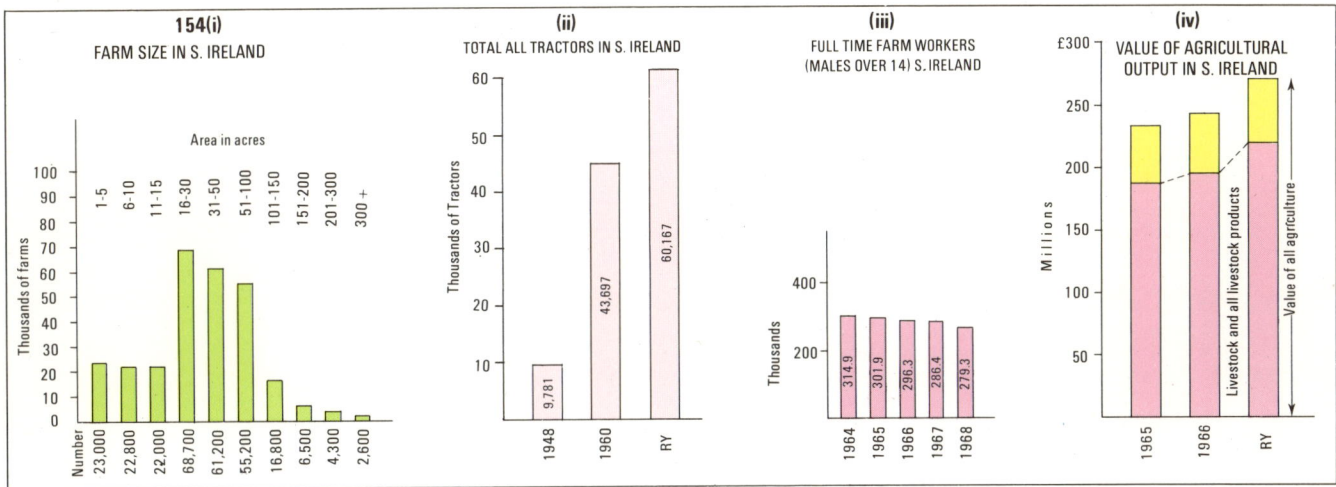

Figs. 153 and 154 Land holdings, mechanisation, labour and agricultural output

153/154a Is Ireland a land of large or small farms?

153/154b Describe the trend in the use of tractors and in employment of labour in both North and South Ireland.

153/154c What branches of agriculture are of increasing importance in both Irelands?

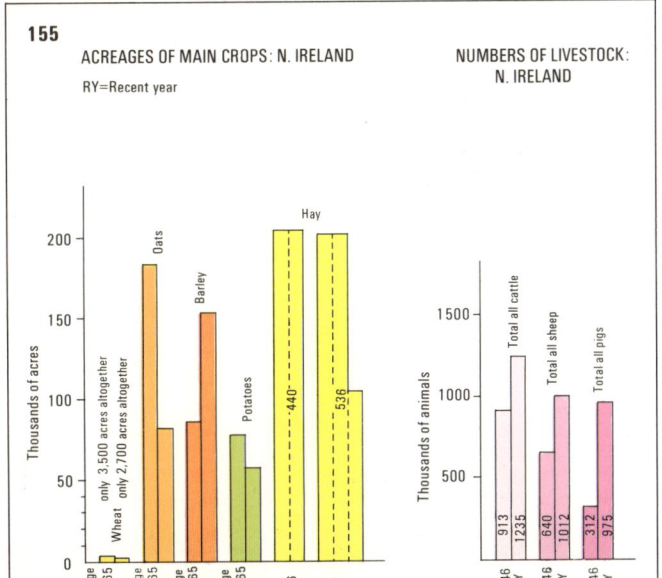

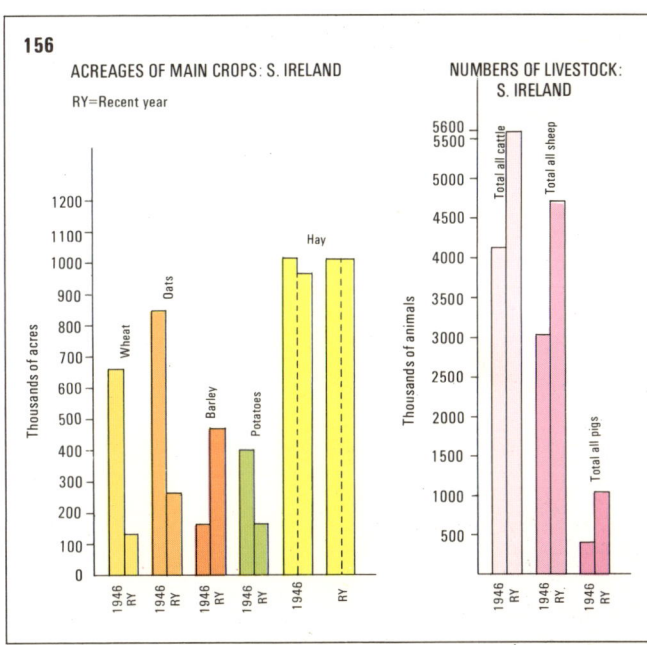

Figs. 155 and 156 Crop and livestock production

155/156a Which cereals have declined in acreage and which has increased in acreage in both Irelands? (That which has increased is used very much as an animal feed today, as well as in the brewing industry.)

155/156b Describe the trends in livestock and hay output for both Irelands.

24 IRELAND Industry and trade

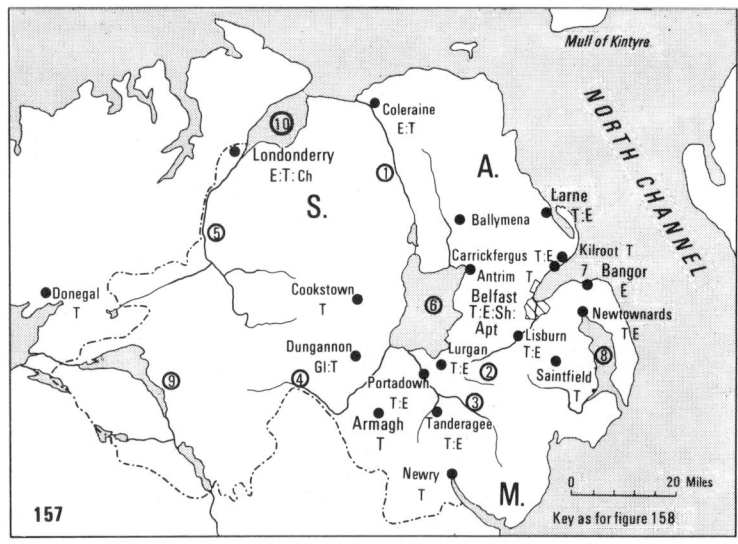

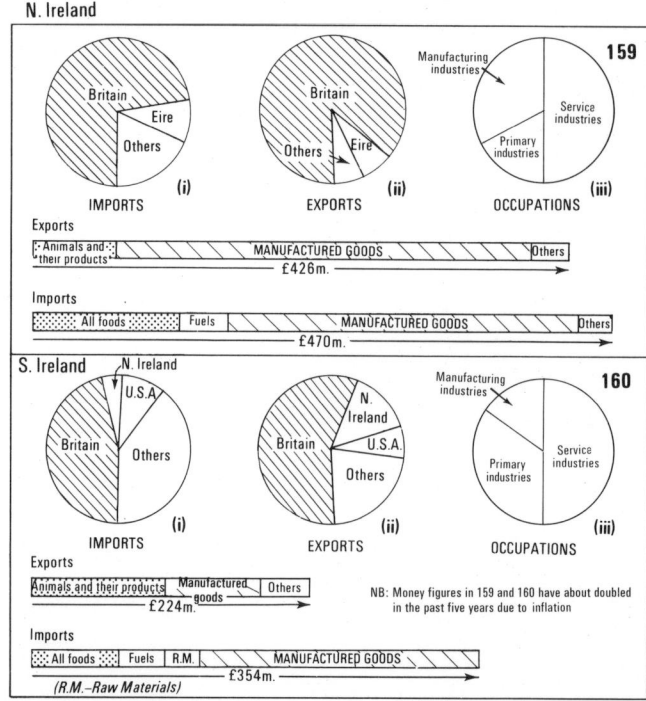

Figs. 157 and 158 Industries and fuels

157/158a For both Northern and Southern Ireland make lists of towns engaged in each of the following industries (a separate list for each industry): textiles, engineering, glassworks, ships, distilling and brewing, chemicals, vehicle manufacturing, sugar beet industry, bacon industry, airport, steelworks.

Note. Local flax and soft water laid the foundations of the textile industry in Northern Ireland. Flax is no longer grown.

157/158b There are seven peat-fired electricity stations in Central Ireland. Name the rivers on which marked H.E.P. stations are situated.

157/158c What climatic conditions favour H.E.P. development in Ireland? (Consider temperatures and rainfall.)

157/158d Name (i) the rivers numbered 1–5 on Fig. 157 and 5–10 on Fig. 158,
(ii) the loughs numbered 6–10 on Fig. 157 and 1–4 on Fig. 158,
(iii) uplands lettered S, A and M on Fig. 157.

157/158e Where does the fuel come from to support the industries of Northern Ireland?

157/158f What advantages does Belfast possess for shipbuilding? (Think of labour supply, coal from Ayrshire and Cumberland, steel from Clydeside and Barrow, and the Lough.)

Fig. 159 and 160 Trade, occupations and foreign investment

159/160a What are the chief exports and imports of both Northern and Southern Ireland?

159/160b How do the graphs show that N. Ireland is more industrialised than Southern Ireland?

159/160c Industrial activity is growing fast in S. Ireland (Eire) with the help of encouraging loans and facilities to foreign investors by the Irish Government. Industrial Income amounted to about £137 mill. out of National Income of about £430 mill. in 1958; by 1967 Industrial Income had risen to about £294 mill. out of N.I. of about £900 mill. Show these figures by two bar-graphs.

159/160d Describe what is indicated by the pie-graphs of Occupations.

159/160e What do the bar-graphs for imports and exports indicate about the fuel resources of Ireland as a whole?

159/160f What do the pie-graphs showing trading partners show about the dependence of Ireland as a whole upon Britain?

159/160g The total capital investment in new industries in Eire, 1959–67, was £109 mill. Examine Fig 160 (ii) and work out the British figure. See Ulster Year Book for 1967–77.

Fig. 161 Communications

161a Name all the places for which initial letters have been marked.

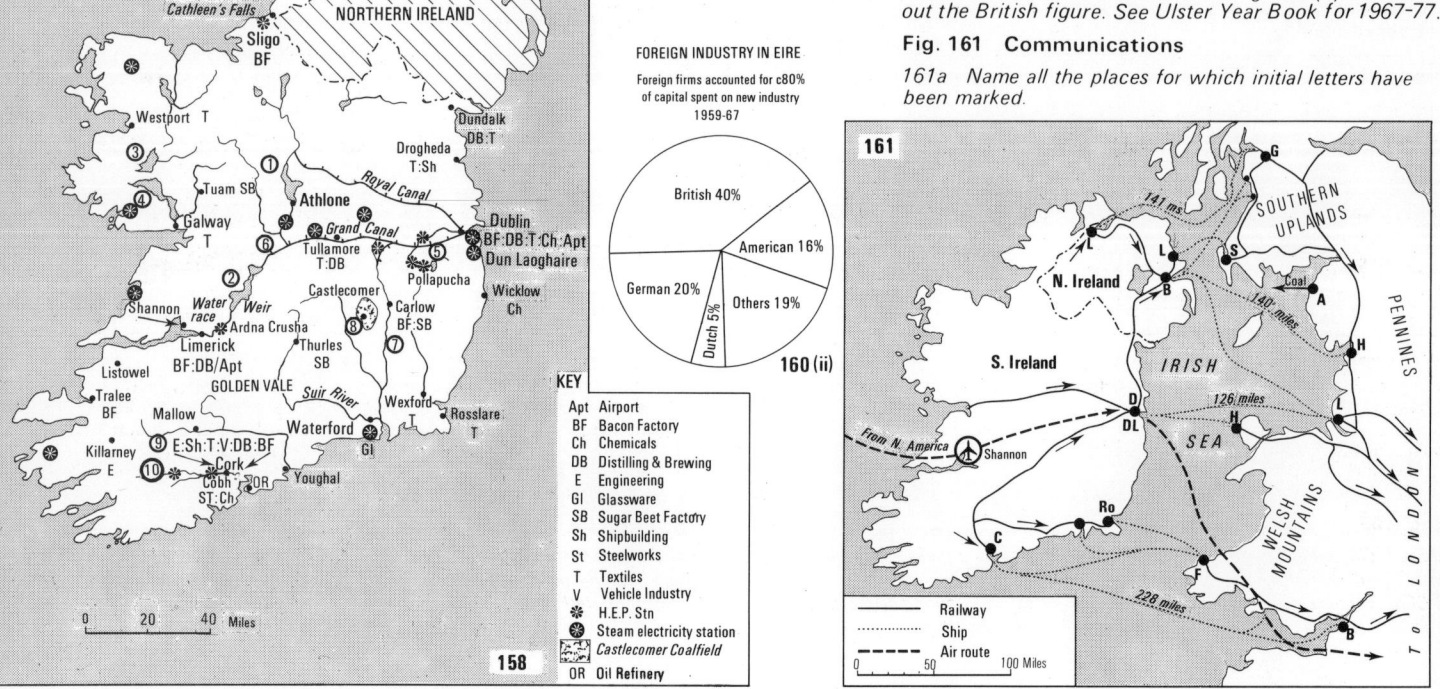

SCOTLAND Regions 25

Scotland can be divided into three geographical sections, on grounds of structure, relief, climate, soils, land use and industrial development, population distribution and occupations. These sections are the Highlands and Islands, the Central Rift Valley, and the Southern Uplands. The distinctiveness of each section will be shown in the following pages.

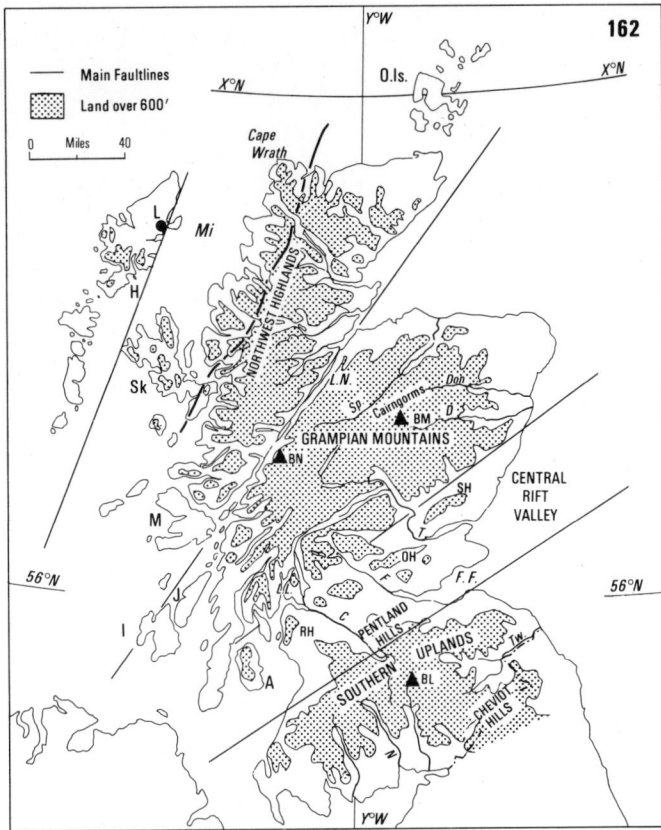

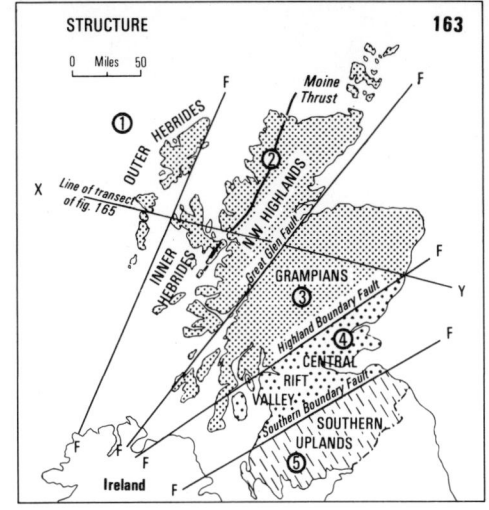

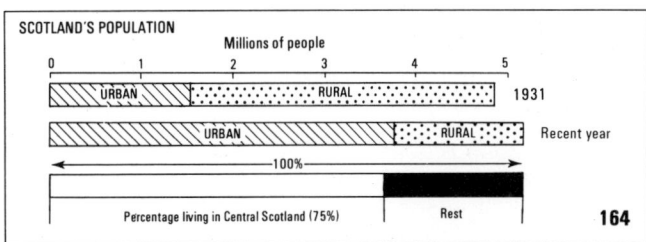

Fig. 162 Relief and drainage

162a Name (i) uplands marked BN, BM, SH, OH, CF, RH, BL,
(ii) rivers marked N, TW, C, F, T, D,
(iii) lochs LN, LL, and water areas Mi, FF,
(iv) islands L and H, Sk, M, J, I, A,
(v) latitude xN and longitude yW.

The three main divisions are shown by different shadings in Fig. 163, which should be consulted together with Fig. 162.

Fig. 163 Structure

The structure of Scotland has been very strongly influenced by faulting, folding and volcanic activity (no longer active). The main fault-lines are shown, and they divide Scotland into five main structural sections. The faulting is of differing kinds, notably tear-faulting, normal faulting, and thrust-faulting.

163a Suggest names for the five structural sections numbered on Fig. 163. (Note: Sections 1, 2 and 3 form the 'Highlands and Islands' region of Scotland.)

163b Find out about, and write an illustrated note on, each type of faulting mentioned above.

Fig. 164 Population

164a Describe the main differences shown by Fig 164 between the population of Scotland in 1931 and today.

164b About how many people live in the Central Valley of Scotland?

Later you will be asked to explain these points about population.

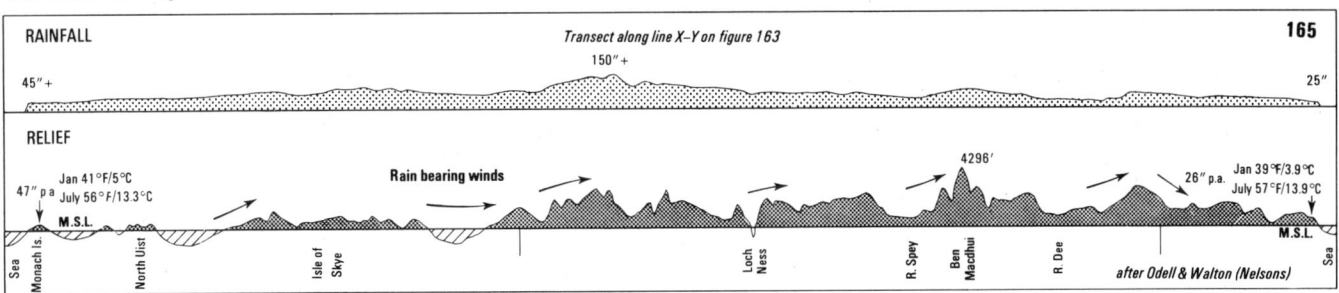

West coast and islands
Deep fjords and fresh-water lochs, formed by valley glaciers. Raised beaches have resulted from an uplift of the land when the ice cap was removed. No coastal plain. Little fertile soil except on deltas and raised beaches.

Interior
Highly glaciated dissected plateau; general level at 3000'. Uplifted peneplain cut into by glaciers during the Ice Age. Fault-controlled valleys common. Very little depth to the rather acidic soil, except in favoured lowland areas.

East Coast
No islands. Straight coastlines. Wider coastal plain. Much glacial deposition—often fertile boulder clay.

Fig. 165

165a Describe the relation between rainfall and relief indicated by Fig. 165.

165b Write a note on (i) the formation of fjords, naming three other parts of the world where fjords may be seen, (ii) features of glacial erosion and deposition.

165c Describe the main differences in the climates of the west and east, indicated by the temperature and rainfall figures given.

26 SCOTLAND Agriculture

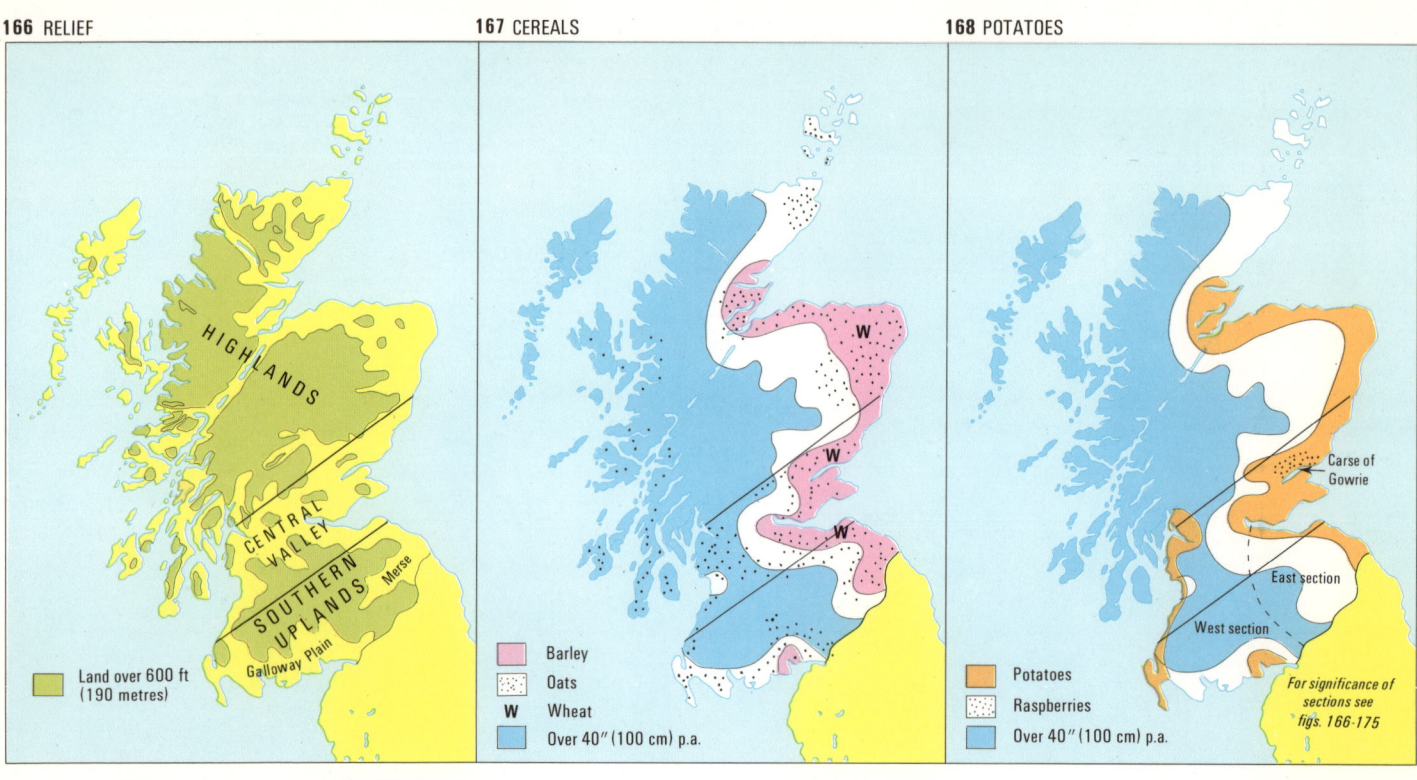

Figs. 166, 167, 168, 169, 170 and 171 should be examined separately and then together.

166/171a Describe the distribution of the main areas of the three cereals in relation to
 (i) relief,
 (ii) rainfall,
 (iii) 'day degrees' (the number of degrees by which temperature exceeds 43°F./6°C. during the year),
 (iv) July sunshine hours,
 (v) length of period free from air frosts.

166/171b What is the length of the frost-free period in (most of) the west coastal strip?

166/171c Describe the influence of the sea and the prevailing winds on the temperatures and rainfall of west Scotland.

Note. The coast of Kent has an average of over 7 hours of sunshine per day during July.

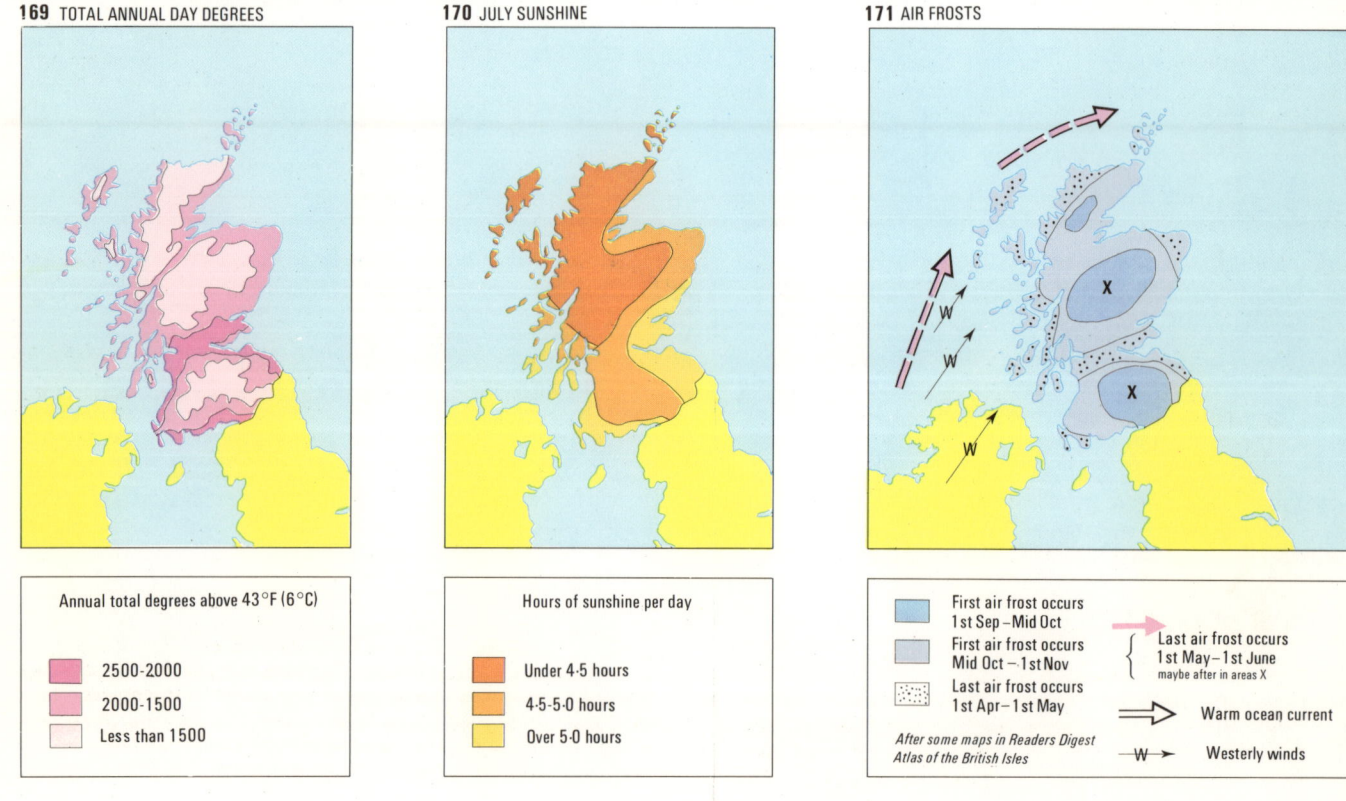

SCOTLAND Agriculture and land use   27

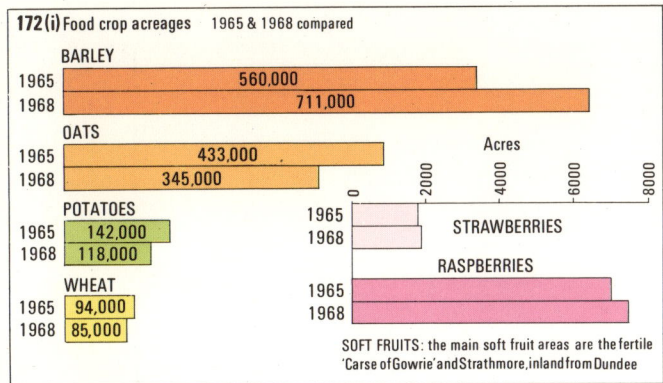

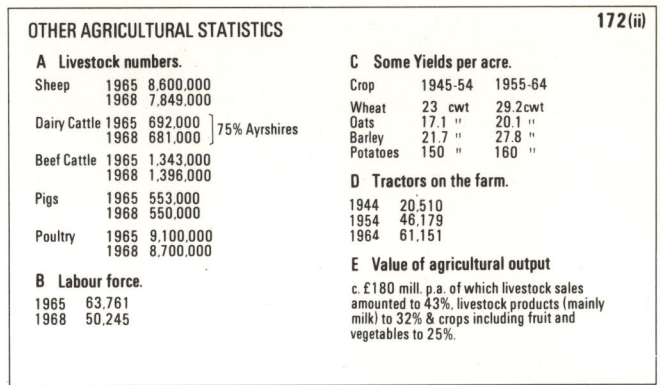

Fig. 172 Crop acreages
Farmers' expectation of future product-prices cause changes in land use and outputs.

172a Describe the changes from 1965 to 1968 in
  (i) cereal acreages, (ii) soft fruit acreages.

172b There are 1,300 'hill sheep farms' in Scotland. These usually average over 4,000 acres, and consist mainly of rough grazing. They are situated in the Highlands and the remoter parts of the Southern Uplands. By how many did the number of sheep increase/decrease from 1965 to 1968?

172c Most of the 7,300 'dairy farms' are in the south-west of Scotland. Each has about 160 acres of crops-and-grass, 70 acres of rough grazing and 42 dairy cows. By how many did the number of dairy cows increase/decrease from 1965 to 1968?

172d Construct a pie-graph to illustrate the information about value of Agricultural Output.

172e Most of the 5,500 'livestock rearing-with-arable farms', and the 2,300 'arable, rearing and feeding farms', are in the NE and E. By how many did the number of beef cattle increase/decrease from 1965 to 1968?

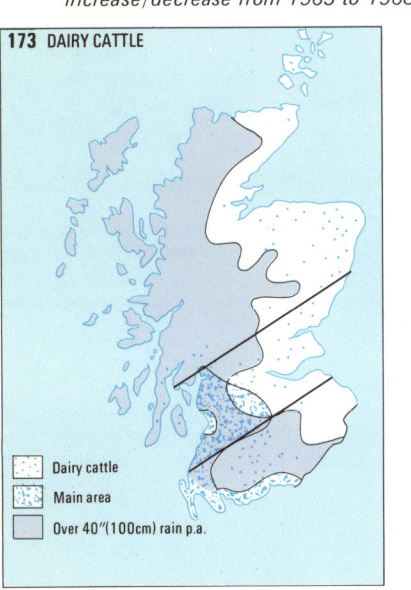
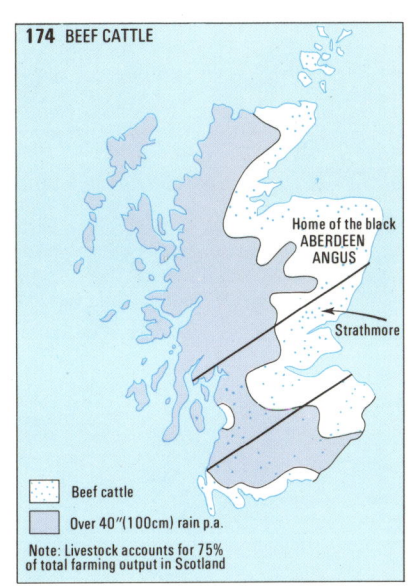
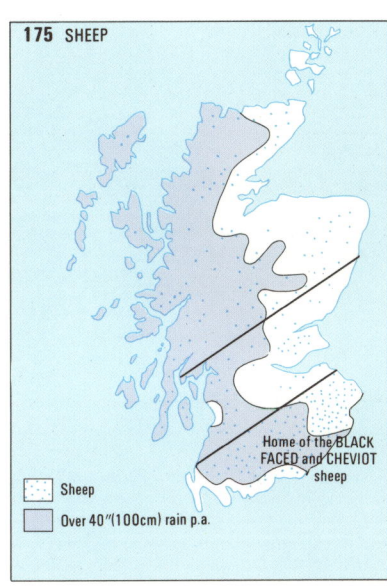

Figs. 173, 174 and 175 Livestock

173/175a Describe the distribution of the three types of livestock illustrated, and give reasons for the concentration of dairy cattle in the western parts of the Central Valley and Southern Uplands.

173/175b Examine Figs. 166–175 and then describe, and try to account for, the major differences in agriculture between Galloway and Dumfries (find these on your atlas) in the west and the Tweed Basin in the east. (Think of crops and livestock, climate, relief.) The rough dividing line between the two sections is shown on Fig. 168.

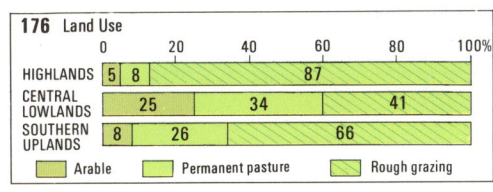

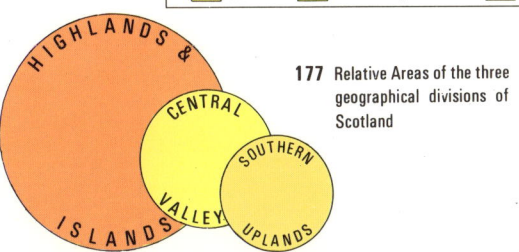

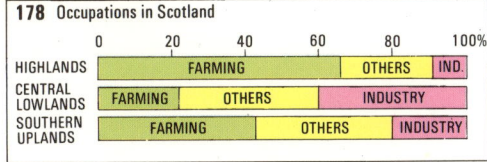

Figs. 176, 177 and 178 Land use and occupations

176/178a What are the main differences between
  (i) the land use in the three divisions of Scotland,
  (ii) the occupations of people of these three divisions.

176/178b Offer possible reasons for the differences in land use.

176/178c Draw circles to represent the areas of the three divisions (Fig. 177) and then, using the percentages shown on Fig. 176, make each into a pie-graph showing land use.

HATFIELD HIGH SCHOOL

28 SCOTLAND Industries and power

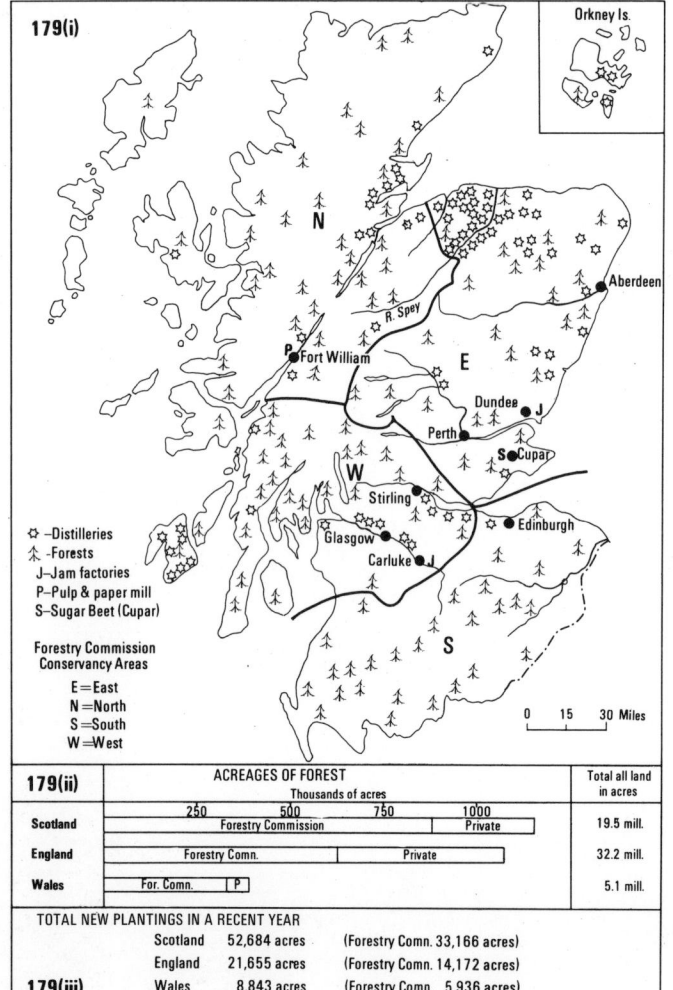

Fig. 179 Industries based on home-grown ingredients

Forestry

The poverty of the soil and the ruggedness of the land surface make farming difficult in much of the Highlands (and S. Uplands). Extensive forests have been planted (mainly Scots pine, Norwegian spruce, larch, fir), and are planned for the future, especially in the Grampians. Remoteness makes development difficult in the north-west.

179a About what percentage of each country is under forest?
179b Describe what is shown by Fig. 179 (iii).

Distilling

P.J. Woodhouse wrote in *The Times*, 23.v.69:

'Scotch whisky is one of Britain's major exports. Shipped to 170 countries, in 1968 it earned more than £176 mill. in foreign currency. The whisky industry has increased its exports by an average of 10%— more than double the rate of British industry as a whole.

There are some 120 distilleries in Scotland. More than 100 of these are malt distilleries. Although most are small they provide work in areas where there is little alternative employment. There are 14 distilleries producing grain whisky. The USA is the largest market, accounting for about half the overseas sales. In 1968 Scotch made up over 10% of the value of all British exports to USA. In recent years . . . the Common Market countries (name them) have been the fastest-growing market for Scotch whisky, with exports rising by over 600% during the past 11 years!

Scotland has several advantages for the production of whisky: it is a local drink, dating from the 14th century or earlier; the use of local peat in the drying process adds a special flavour to the drink; and the quality of the water, e.g. Spey valley (39 distilleries), is excellent for the purpose. Railway development was of great importance for marketing (see Speyside).'

179c Name the western island and the area of the mainland which are most important for whisky production.

Fig. 180 Hydro-electric power in the Highlands

The Scottish Highlands are not really ideal for the production of H.E.P. because the natural catchment areas are too small. This difficulty is being overcome by the diversion of rivers to combine formerly separate river systems.

180a What advantages do the Highlands have for H.E.P. production?

Hugh Cochrane wrote in *The Times*, 5.viii.68:

'The North of Scotland Hydro-Electric Board's area is now 21,638 sq. miles, north and west of a line from the Firth of Clyde to the Firth of Tay. There are cities like Aberdeen and Dundee within it, but more common are broad stretches of farmland, lonely glens and crofting townships, and the islands. Average consumer density is not usually more than 21 per sq. mile, against 178 in the rest of Scotland, and 301 in England and Wales. In the past 20 years, 50 major dams and 53 main power stations have been built; nearly 200 miles of rock tunnel excavated, and about the same length of aqueducts and pipelines laid; 400 miles of roads have been laid or re-laid, more than 20,000 miles of overhead cable strung, and about 450 homes built for staff in remote places. *The pace of expansion has been heavily curtailed in recent years . . . the burden of interest and financing expenses have become heavier; the need to prove (the) viability of new works and to improve efficiency has become sharper.* To date the Board have connected 96% of the potential consumers in their region. In 1949 . . . there were just over 200,000 consumers, providing an annual revenue of about £1·8 mill. In the latest financial year there were more than 444,000, with a total revenue of about £28 million.'

180b Explain the meaning of the italicised sentence above.
180c What effect do you think this expansion in electrical provision in the Highlands will have, and is having, on the depopulation of the region?
180d Make a list of all the H.E.P. stations marked on the map (i) in the N.W. Highlands, (ii) in the Grampians.

The aluminium industry (controlled by the British Aluminium Company)

The bauxite from Ghana and France enters at Burntisland, where it is purified to aluminium oxide. This oxide (alumina) is sent to be reduced to aluminium, using H.E.P. at Lochaber and Kinlochleven. Super-fine aluminium is purified further at Foyers. The purified aluminium then goes by road to Falkirk rolling mills.

180e On a map of Scotland mark the sites and movement described.

The Tourist industry

Good publicity and improved roads have helped the tourist industry to increase greatly in the Highlands.

180f What are the tourist attractions in the Highlands?

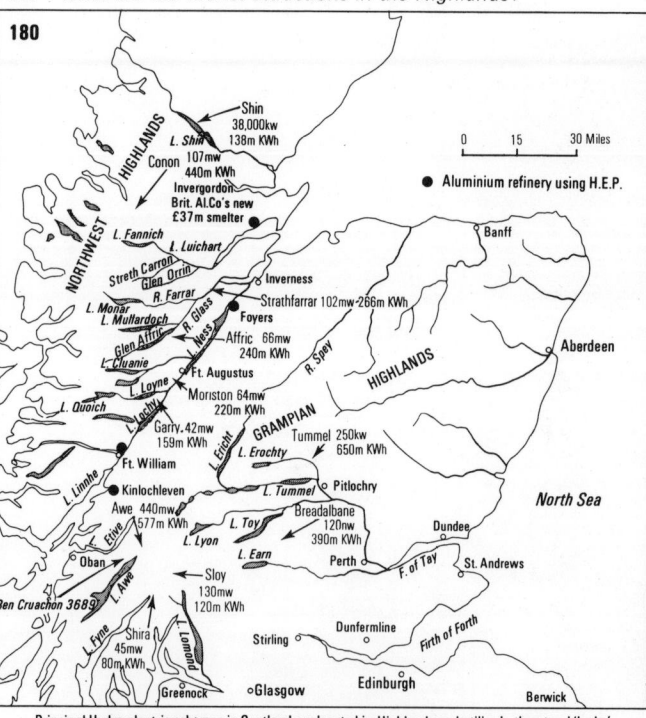

Principal Hydro-electric schemes in Scotland are located in Highlands and utilize both natural 'lochs' and a few man-made ones (Map from article by Waldo Bowman, Editor of Engineering News-Record)

SCOTLAND   Industry in the Central Lowlands and Southern Uplands

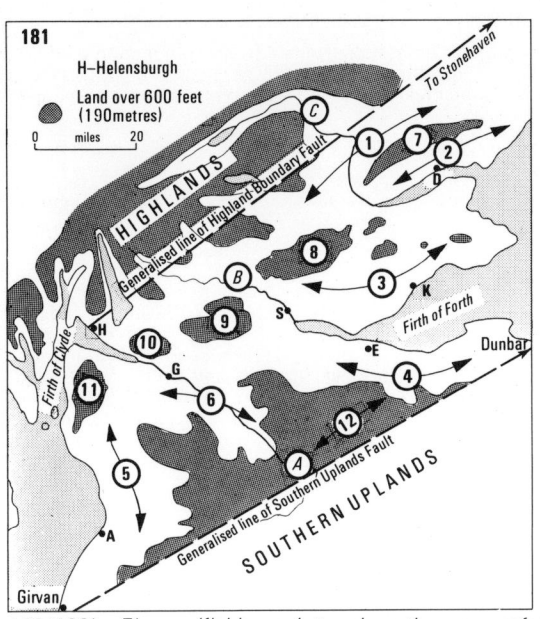

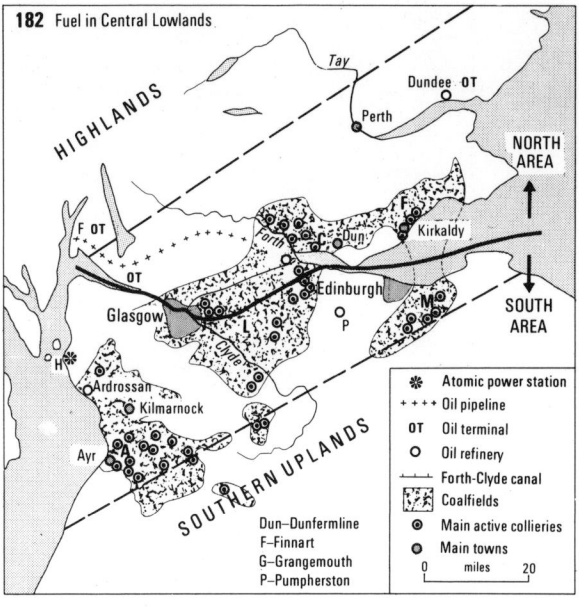

Fig. 181 The Central Lowlands

181a Examine Fig. 181 and then name
(i) lowlands 1–6,
(ii) uplands 7–12,
(iii) rivers A, B and C,
(iv) towns D, G, S, K, E.

The hills 7–12, which have all resulted from the upwelling of igneous rocks near the boundary faults, have thin soils and adverse climatic conditions, and are grazed by hill sheep. They contrast greatly with the rich intensively-farmed fertile lowlands surrounding them.

Figs. 182 and 183 Industrial development depends upon power supplies, among other things.

182/183a Name and locate the different sources of industrial power in the Central Lowlands and Southern Scotland.

182/183b The coalfields are lettered on the map as follows: A, Ayrshire; L, Lanark; C, Clackmannan; F, Fife; M, Midlothian. Name the atomic power station situated in Central Scotland.

182/183c Between which coalfields does the River Forth run?

(Figures of coal output are shown on page 30.)

182/183d List the towns engaged in
(i) the steel industry,
(ii) the aluminium industry.

Ravenscraig has a new hot strip mill. The plant is designed to produce 12 million tons of steel strip per year, and supplies steel sheet for the new car body factory at **Linwood** (near Paisley)—output capacity is 4,000 car bodies per week, using 2,000 tons of steel sheet. Initial advantages of the area for the iron and steel industry were:

(i) the presence of local blackband iron ore (now exhausted),
(ii) hard splint coal for smelting,
(iii) plentiful water supplies,
(iv) the introduction of the Neilson hot blast system in 1828.

Shipbuilding, engineering and textiles, and now the expanding new light industries, provide a ready market for the industry's products, and 15% of the output is exported.

(Figures of the output of the iron and steel industry are given on page 30.)

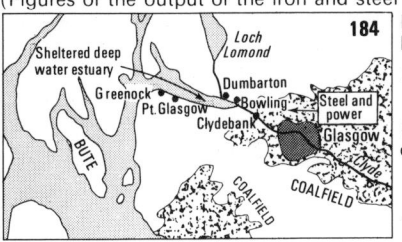

Fig. 184 Shipbuilding and Marine Engineering on Clydeside

184a What were the natural advantages of Clydeside for shipbuilding?

(Figures of shipping production are given on page 30.)

Fig. 185 Textiles and power production

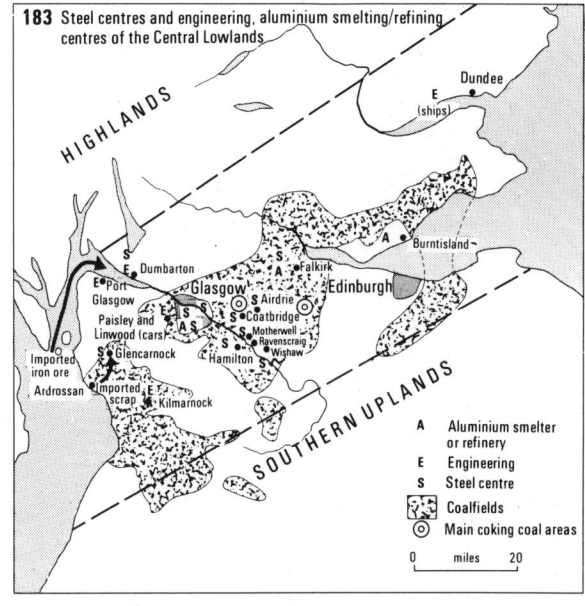

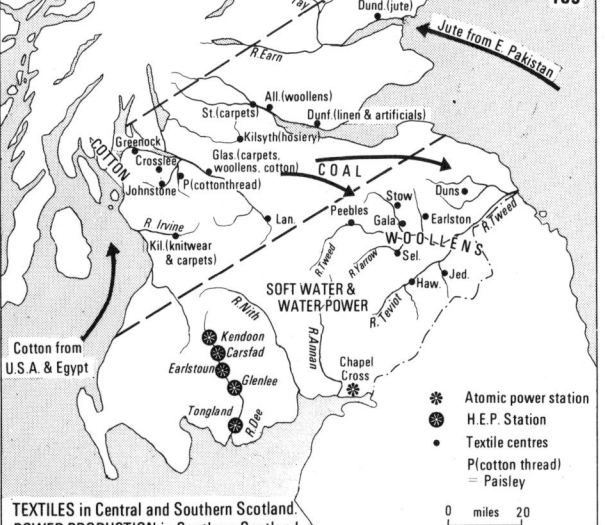

185a What common site factor have all the textile centres shown on Fig. 185?

The textile industry in the Central Lowlands was encouraged by
(i) the local wool supplies,
(ii) the skills of the people, developed over the centuries,
(iii) the plentiful supply of water for washing, bleaching and dyeing, and for motive power,
(iv) the Trade Triangle of the 18th century, which led to the import from America of tobacco, sugar cane and *cotton*,
(v) the damp climate.

185b What was the Trade Triangle?

185c How was point (v) an advantage?

185d Why should a growing wealth among the increasing population of the Central Valley itself, and the opening up of India and former British lands in Africa, have helped the textile industry?

185e The American Civil War (dates?) was a serious blow to the cotton textile industry. Why?

185f Figs. 175 and 185 give clues as to the advantages of the Tweed Valley and its tributaries for the woollen industry there. What are they?

Today the eastern valley towns of the S. Uplands specialise in high quality garments, and rayon has been introduced in Jedburgh.

185h What is the town (not named on Fig. 185) at the mouth of the Tweed?

185i Name the textile towns which have been shown on Fig. 185 by only part of their full names.

185j What influence has the widespread use of
(i) paper sacks, (ii) linoleum and cheap carpets
had on the jute industry at Dundee?

SCOTLAND Coal, iron, steel, ships

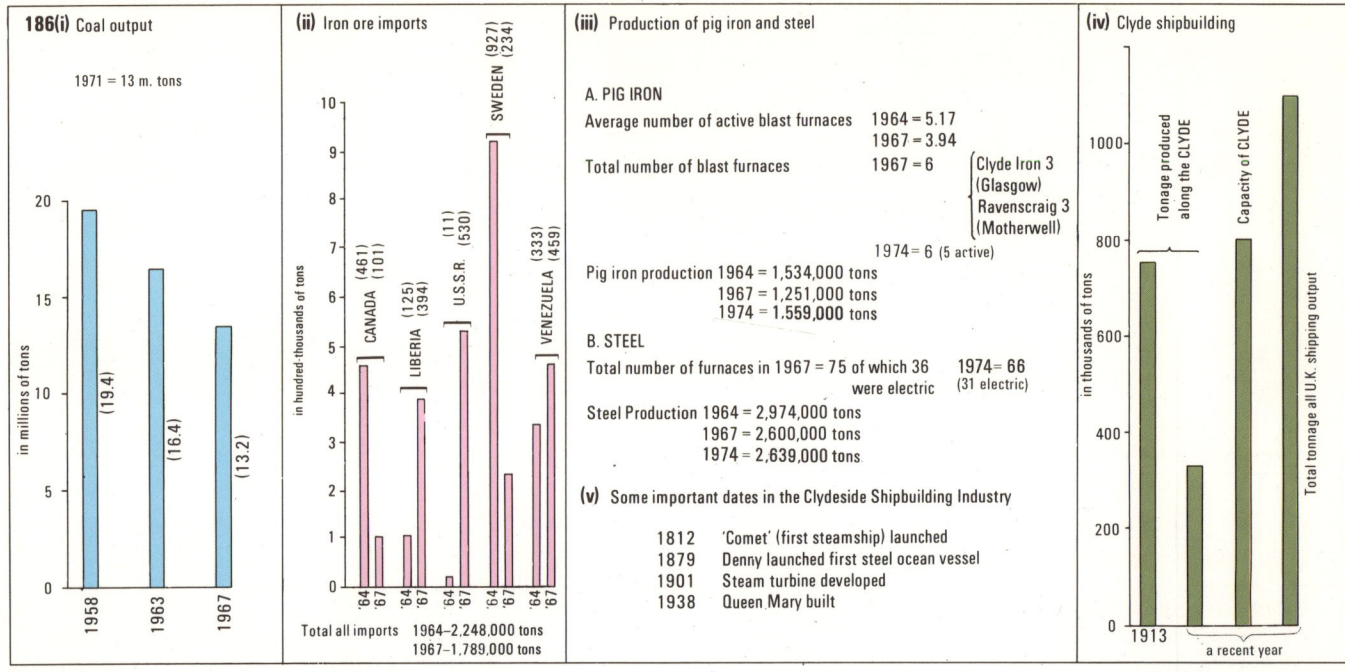

186(i) Coal output 1971 = 13 m. tons
(1958 = 19.4; 1963 = 16.4; 1967 = 13.2, in millions of tons)

(ii) Iron ore imports (in hundred-thousands of tons), 1964 and 1967:
- CANADA (461)(101)
- LIBERIA (125)(394)
- U.S.S.R. (11)(530)
- SWEDEN (927)(234)
- VENEZUELA (333)(459)

Total all imports 1964 – 2,248,000 tons; 1967 – 1,789,000 tons

(iii) Production of pig iron and steel

A. PIG IRON
Average number of active blast furnaces 1964 = 5.17 ; 1967 = 3.94
Total number of blast furnaces 1967 = 6 { Clyde Iron 3 (Glasgow), Ravenscraig 3 (Motherwell) }
1974 = 6 (5 active)
Pig iron production 1964 = 1,534,000 tons ; 1967 = 1,251,000 tons ; 1974 = 1,559,000 tons

B. STEEL
Total number of furnaces in 1967 = 75 of which 36 were electric   1974 = 66 (31 electric)
Steel Production 1964 = 2,974,000 tons ; 1967 = 2,600,000 tons ; 1974 = 2,639,000 tons

(v) Some important dates in the Clydeside Shipbuilding Industry
- 1812 'Comet' (first steamship) launched
- 1879 Denny launched first steel ocean vessel
- 1901 Steam turbine developed
- 1938 Queen Mary built

(iv) Clyde shipbuilding (in thousands of tons): 1913 – Tonnage produced along the Clyde ~750; Capacity of Clyde ~325. A recent year – Capacity of Clyde ~800; Total tonnage all U.K. shipping output ~1075.

The four basic industries in the Central Valley of Scotland have been coalmining, iron and steel manufacturing, shipbuilding and textiles. In textiles the tweed industry is neither expanding nor contracting, the cotton industry is contracting, and the knitwear industry is expanding.

Fig. 186 Coal output, iron ore import, iron and steel production

186a What are the trends in the other major industries?
186b Describe the changes in imports of iron ore as shown in Fig. 186.
186c Draw a bar-graph to illustrate steel furnace figures for 1967.
186d When and where were the Queen Elizabeth and the Q.E. II built?
186e Approximately what fraction of total UK shipping tonnage was produced on the Clyde?
186f How much below capacity was the output of shipping tonnage on the Clyde?

The decline in basic industries in Scotland has led to the introduction of Government-backed schemes for new developments, especially of light industries such as aero-engines, cars, shavers, domestic appliances.

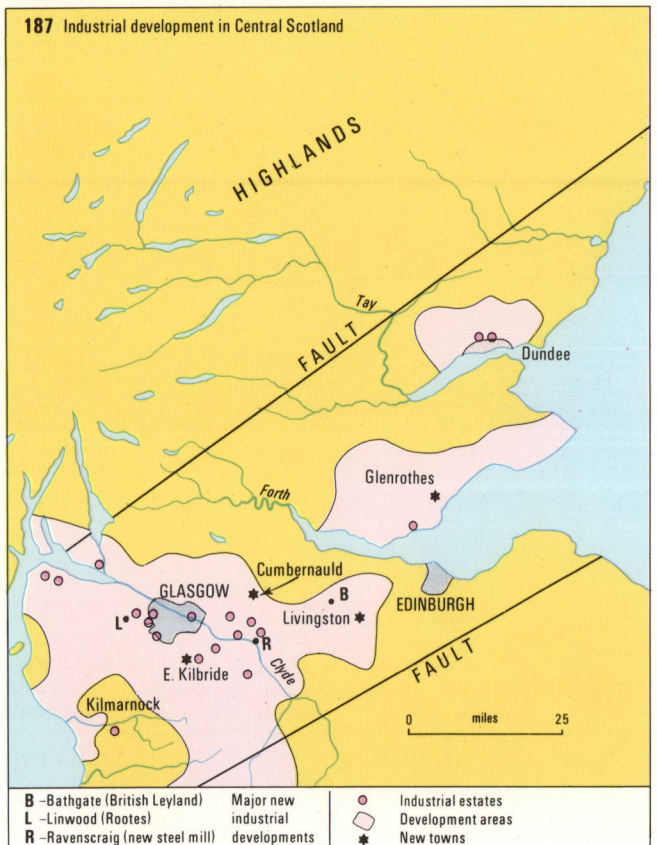

187 Industrial development in Central Scotland

B – Bathgate (British Leyland)
L – Linwood (Rootes)
R – Ravenscraig (new steel mill)
Major new industrial developments
○ Industrial estates
Development areas
✶ New towns

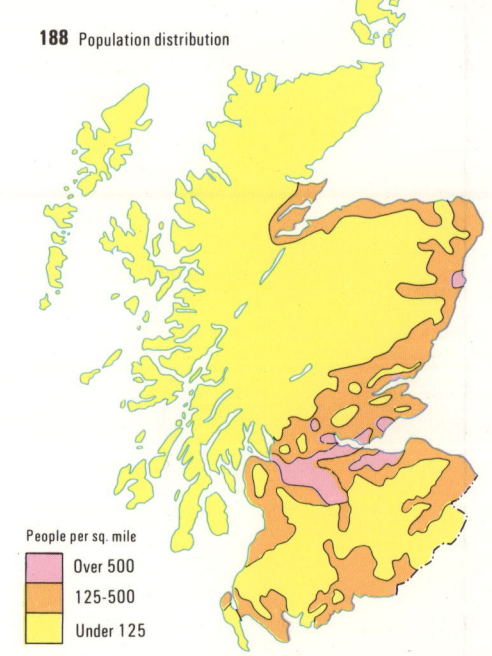

188 Population distribution

People per sq. mile:
- Over 500
- 125–500
- Under 125

Fig. 187 Industrial development in Central Scotland

187a Suggest reasons for the development of new towns to house the people, especially from Glasgow.

Fig. 188 Population distribution in Scotland

188a Describe and give reasons for the distribution of population. (This requires you to explain areas of sparse population as well as those which are heavily populated.)

ENGLAND   Lake District and surrounding plains   31

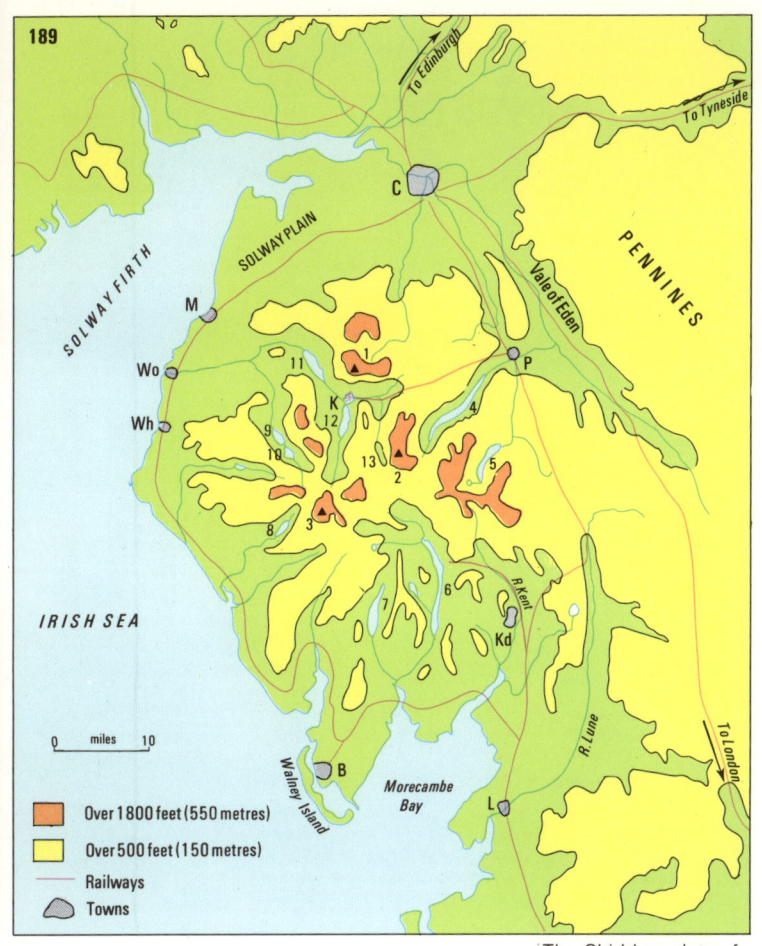

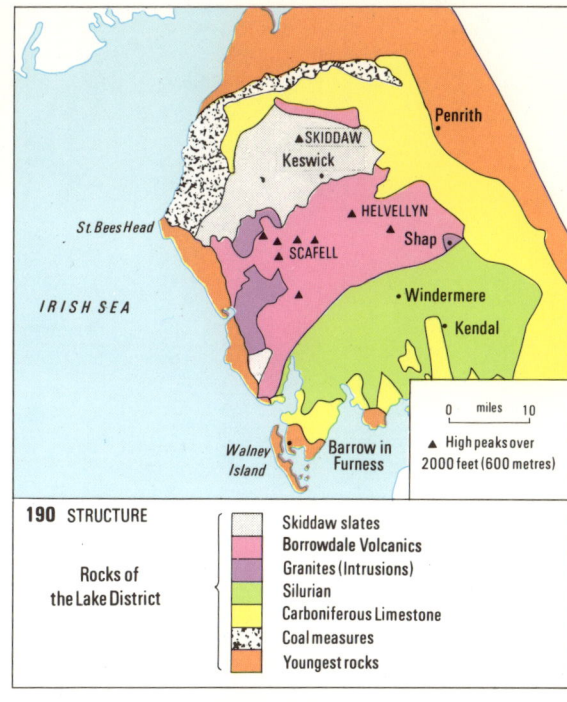

Fig. 189 Relief and drainage

189a Examine Fig. 189 and then name
(i) mountains numbered 1–3,
(ii) lakes numbered 4–13,
(iii) towns marked M, Wo, Wh, C, Kd, K, B, P, L.

189b Describe the pattern of drainage shown.

Figs. 190 and 191 Structure

190a Note the positions of Keswick, Penrith, Kendal and Scafell Peak on Figs. 189 and 190.

The Skiddaw slates form smoothed, rounded uplands, while the Borrowdale volcanics provide rugged peaks and generally 'wild' scenery. The whole area has been glaciated—mountains vigorously eroded, and boulder clay smeared over the valley floors and lowland areas.

190/191a Relate the section (Fig. 191) to the map (Fig. 190).

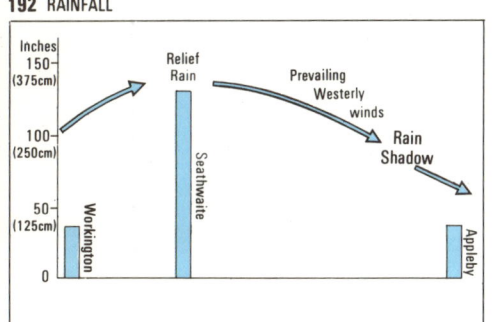

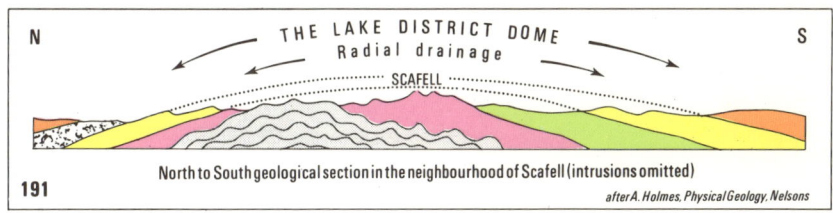

Figs. 192 and 193 Climate

192/193a Describe and account for the differences in rainfall.

192/193b How many months does Keswick experience with temperatures below 43°F./6°C.?

192/193c Describe and account for the contrasts in the total sunshine hours of Keswick and Dover shown on Fig. 193.

192/193d Pick out TWO words from the following which best describe the climate of the interior of the Lake District: cold, hot, mild, wet, dampish, dry.

192/193e What disadvantages of climate and relief does the Lake District suffer as far as wheat growing is concerned? (Examine Figs. 189, 192 and 193.)

Fig. 194 Transect showing agricultural land use

The heavy rainfall, the mild temperatures throughout the year, the lack of rich flat land suitable for arable farming, and the abundance of rough pasture on the fells have all combined to encourage pastoral farming in the Lake District. The dairy and the store cattle are concentrated in the valleys, which are drained and limed for pasture and fodder crops, while the sheep spend the summers in the fells and the winters in the valleys or surrounding lowlands (Solway Plain). The Vale of Eden is a clay-covered lowland devoted to mixed farming, with dairying dominating. The Forestry Commission has planted coniferous trees on slopes. There are now some 13 forests in the north-west.

194a Describe the transect.

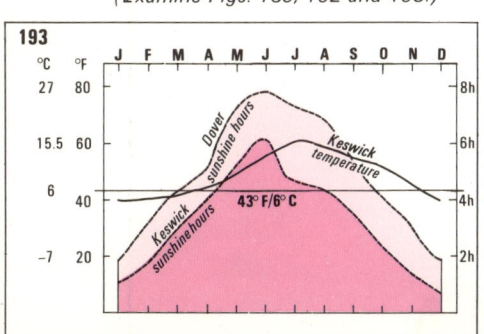

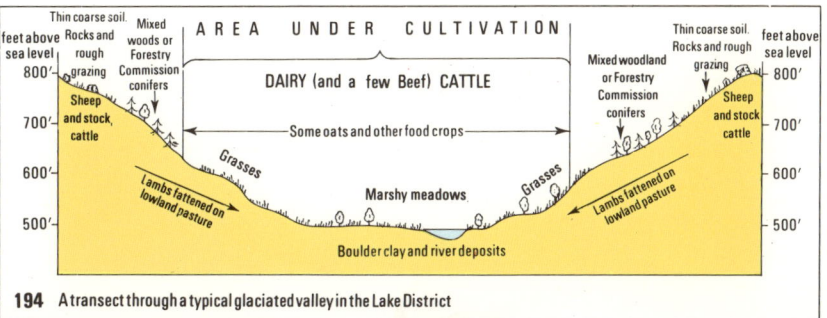

ENGLAND Lake District and surrounding plains

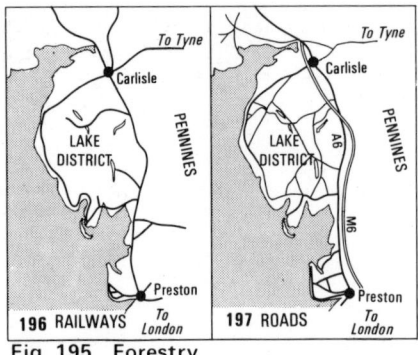

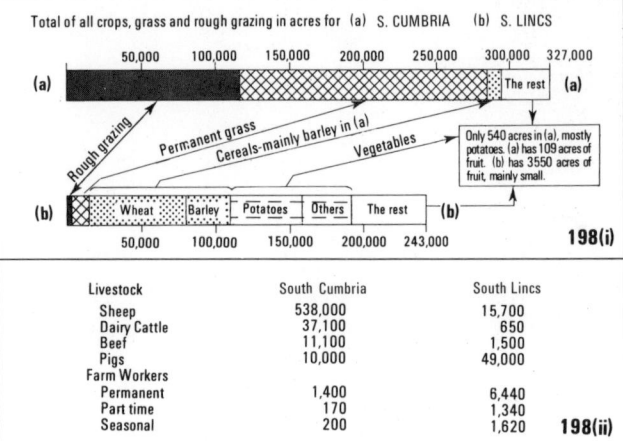

Fig. 195 Forestry

195a Examine Figs. 194 and 195. Name the forest plantations of over 5,000 acres.

Figs. 196 and 197 Communications

196/197a What influence has Lake District relief upon the road and rail pattern?

196/197b Two roads in the region are of special importance. Which are they, and why are they so significant?

196/197c Suggest the influence of the new M6 on the Lakeland tourist industry.

Fig. 198 Agricultural output

198a Study the output figures. Mention, and try to account for, the great contrasts in (i) crops, (ii) livestock farming between S. Cumbria and S. Lincs. (Think of climate, relief, soil.)

198b Account for the large number of seasonal workers in the old Holland County.

Fig. 199 Minerals and power resources

199a Describe the locations of (i) the Cumbrian coalfield, (ii) the iron ore at present worked, (iii) roofing slate, (iv) road granite, (v) atomic power stations.

199b The possible site of a barrage across Morecambe Bay is shown on Fig. 199. Find out and write about the Barrage Scheme.

199c The Lakeland tourist industry is expanding. Why?

199d What local raw materials have encouraged the development of a chemical industry at St Bee's head?

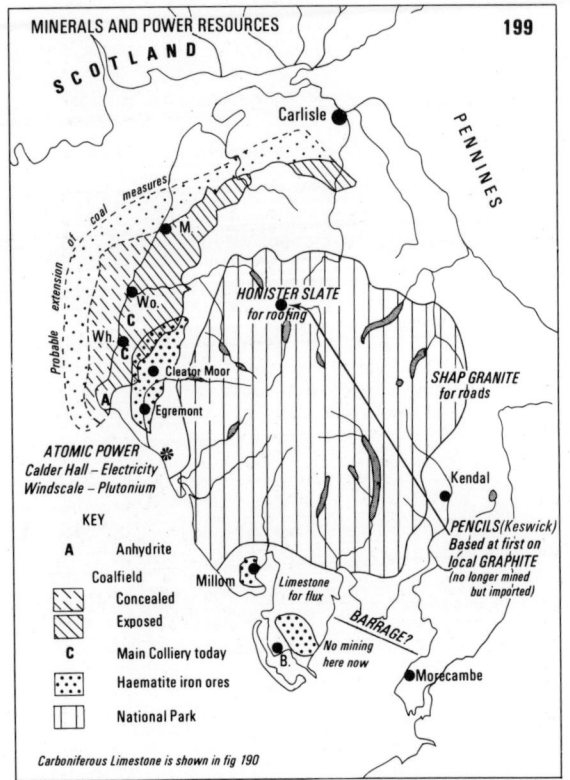

Fig. 200 The iron and steel industry

This industry was encouraged by local iron ore (haematite, 49% pure), Cumbrian coal, and Furness district limestone (for flux).

200a Is the iron and steel industry expanding or contracting? (Illustrate your answer with figures.)

200b The main imports of iron ore today come from three countries. Name them and describe recent changes in their importance to the north-west of England.

200c Note the numbers of blast and steel furnaces. Is the coal of Cumbria (or Durham) of very much direct importance in the industry today? Explain.

200d Describe the recent changes in the coal industry of the much-faulted Cumbrian coalfield (give figures).

Some of the steel is used in the SHIPBUILDING industry of Barrow-in-Furness, though most of the steel plate comes in from Scunthorpe (Lincs.) and the Clyde Valley (Scotland). Today Vickers-Armstrong concentrates on building submarines and tankers, but almost every type of ship has been built there.

200e Why has Walney Island been of importance to the development of shipbuilding at Barrow?

Periods of unemployment have characterised the shipbuilding industry; the iron and steel industry is declining; and the Government has made this a Development Area. This means that cheap subsidised rents are asked for factories, and generous grants are offered to firms who will operate within the Area. New light industries have therefore developed, e.g. at Workington: Ectona Fibres, 1968, which give employment not only to men but also to their wives and daughters.

200f As the steel industry has closed down there has been a tendency for families to move away. What influence does this have on the new industries, which employ labour in the ratio of 2 females to 1 male?

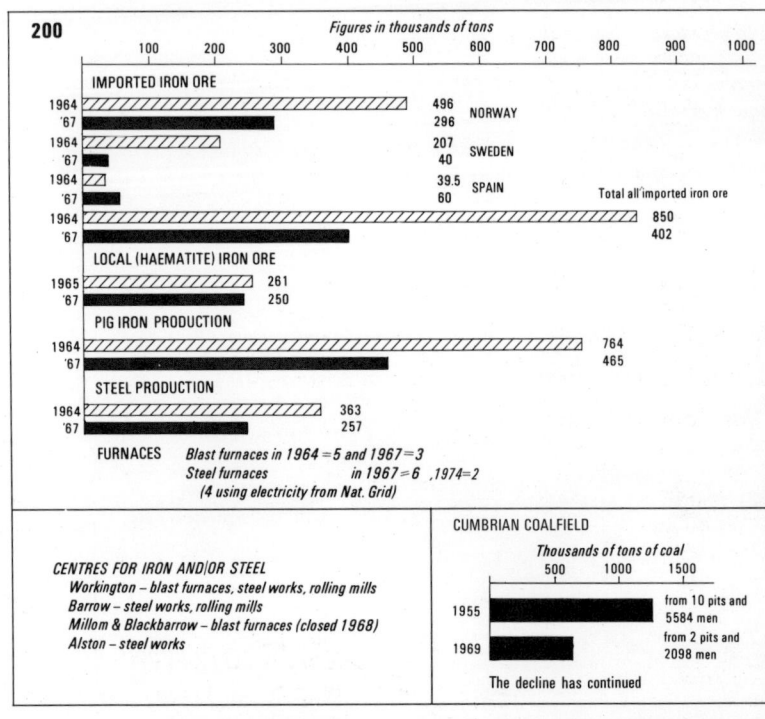

ENGLAND   The Pennine Hills

RELIEF, STRUCTURE, DRAINAGE AND LIMESTONE SCENERY

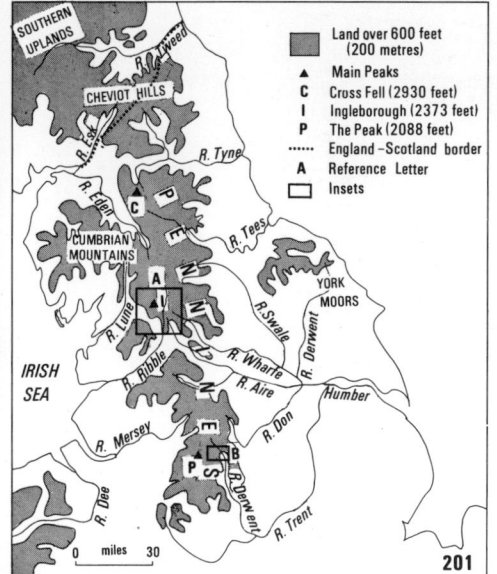

Fig. 201   Relief and drainage

201a What hills form the northern limit, and what river forms the southern limit of the Pennine range?

201b Calculate the approximate length (in miles) of the Pennines.

201c Work out the average height of the three highest peaks.

201d Distinguish two east-west gaps in the Pennines and associate them with two rivers. (The river Tees and Eden also form a less conspicuous gap called the Stainmore Gap.)

These gaps provide important through-routes for roads and railways. The M62 motorway linking Lancashire and Yorkshire will be completed in 1971.

202a What two main rocks make up the Pennines, and in what parts do they occur?

202b On the flanks of the Pennines exist large areas of a rock valuable to man. What is it?

202c Name the coalfields marked (initials given).

Fig. 203   Cross-sections (for lines of cross-sections, see Fig. 202).

203a Make a list of the succession of rocks in the Pennines from oldest to youngest.

203b The Central and Southern Pennines have been folded into a structure called an anticline. What do you understand by this term?

203c The structure of the Northern Pennines differs from that of the Central and Southern Pennines. Explain this difference, using the following words and phrases: fault zone, Eden Valley, downfaulted, fault scarp dipping gently to the east, anticline.

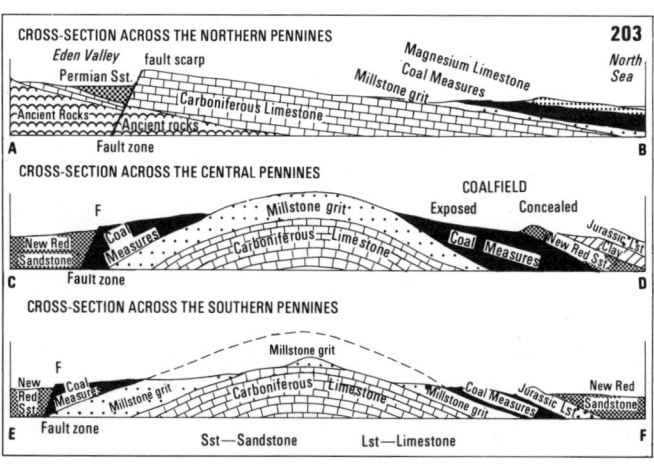

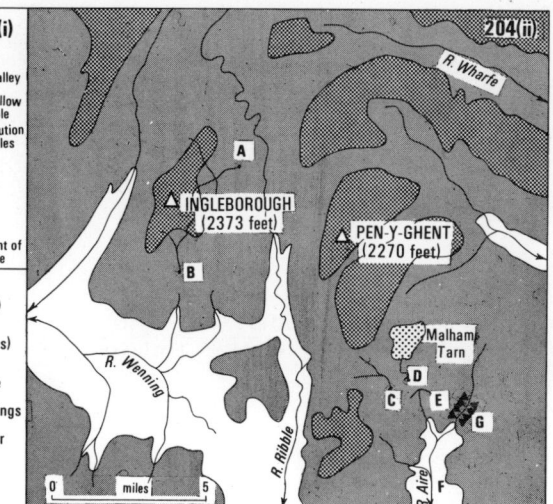

Fig. 204   Limestone scenery (See inset A Fig. 201.)

204a The Ingleborough area of the Pennines exhibits the unique surface features of limestone scenery. Using a physical geography book to help you, name, explain, and give examples from this area of, four features of limestone scenery.

AGRICULTURE, WATER SUPPLY, INDUSTRY AND TOURISM

Fig. 205   The Pennines west of Sheffield (See inset B, Fig. 201.)

205a In what ways does farming differ in the Limestone and Millstone Grit areas of the Pennines?

205b What factors favour dairying and the cultivation of fodder crops in Edale? (Think of temperature, rainfall, wind, soil, slope, etc.)

205c Most of the milk produced in the Pennines is sent to the large urban areas to the east and west. Suggest four such urban areas. Some of the milk is converted into cheese, e.g. Wensleydale.

205d Annual rainfall in the high Pennines exceeds 60 inches. What other factor favours the development of reservoirs in some parts of the Pennines? (Consider rock types.)

205e Name two large towns likely to receive water from the Ladybower Reservoir.

205f Distinguish two Pennine industries—one past and one present.

205g Tourism is also an important industry. The Pennines possess the same attractiveness as the Lake District. Explain this.

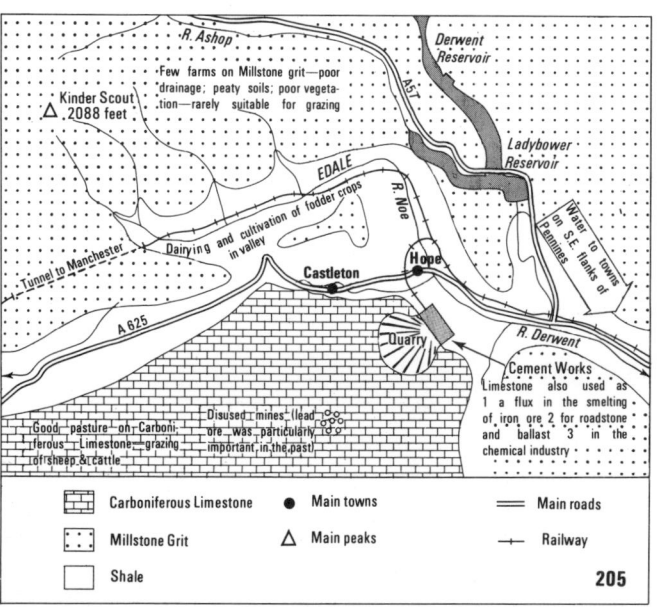

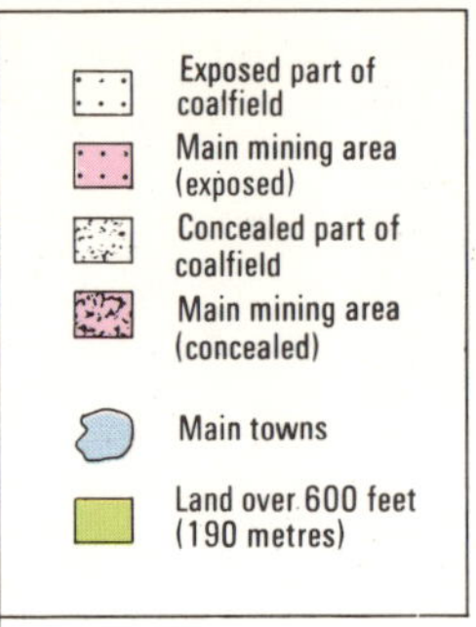

34 ENGLAND The Yorks., Notts. and Derby Coalfield

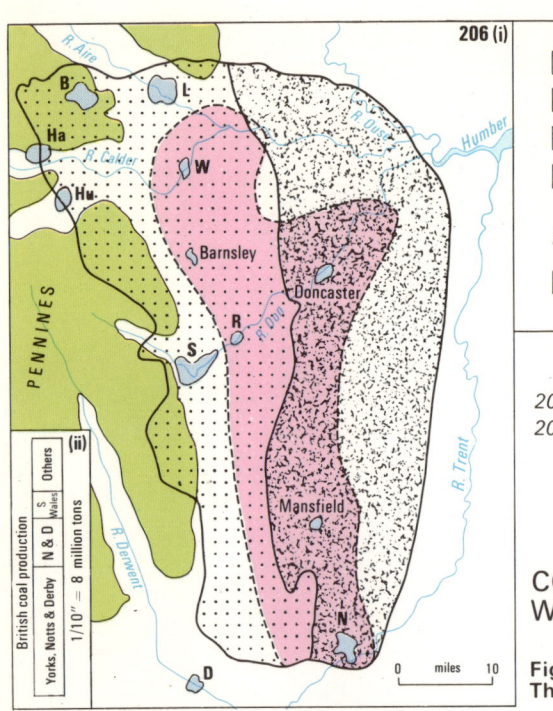

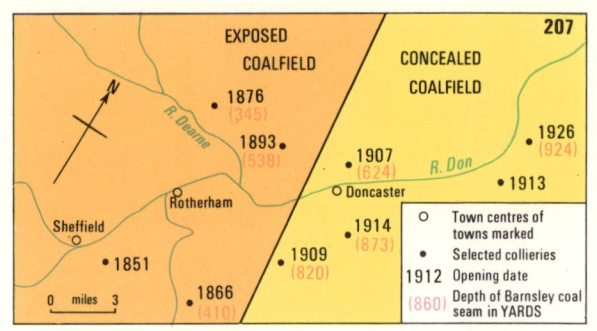

Fig. 207 Collieries in Sheffield/Doncaster area

207a Are the newest mines on the exposed or the concealed part of the coalfield?
207b (i) What happens to the depth of the Barnsley seam as you go eastwards?
       (ii) How will this affect the working of the collieries in the eastern part of the coalfield?

COALMINING, WOOL TEXTILES

Fig. 206. The coalfield

206a What are (i) the approximate dimensions (north-south, east-west), and
             (ii) the approximate northern and southern boundaries of this coalfield?
206b The main coalmining towns are named in full. List them and describe the position of each.
206c Why are the terms 'exposed' and 'concealed' used? (See Fig. 203.)
206d Is the western or eastern part of this coalfield exposed?
206e What percentage of total British coal production comes from this coalfield?

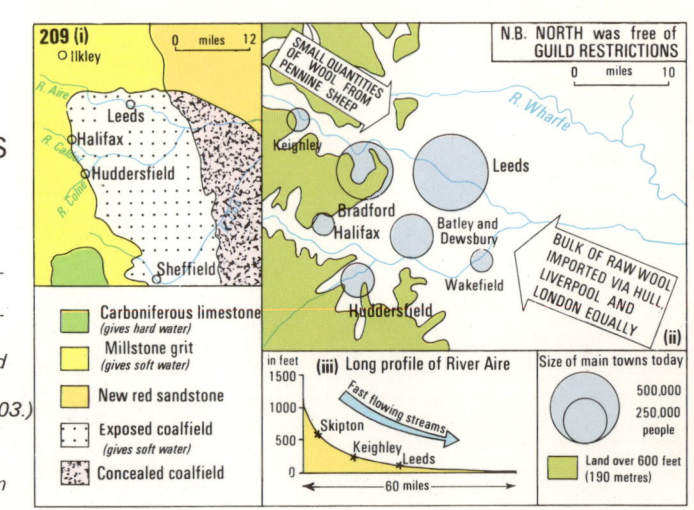

Fig. 208 Industrial employment in Yorkshire and Humberside, recent year

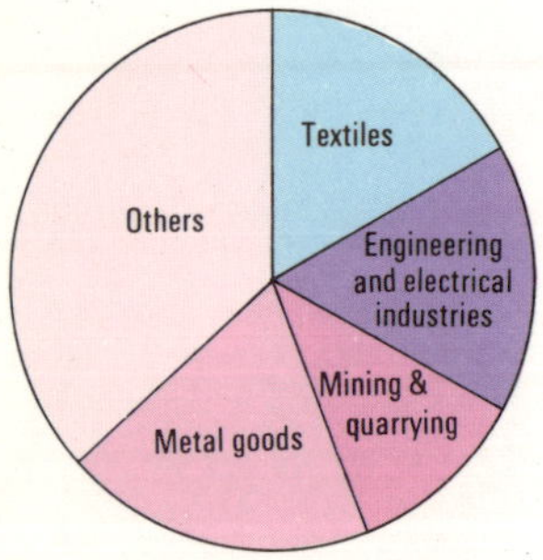

208 Total industrial employees: 1,024,000

This region is the northern half of the coalfield and the area immediately north and south of the Humber estuary.

208a What are the four main industries in Yorkshire and Humberside?

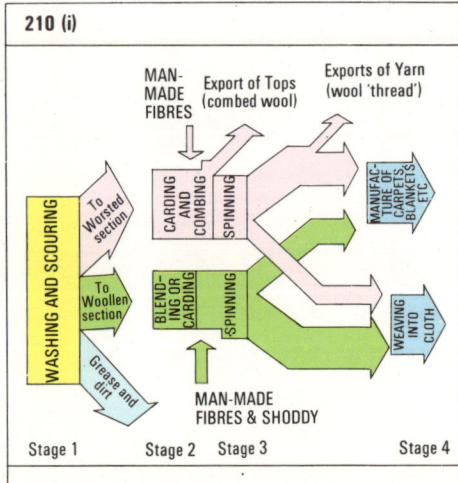

The width of the flow lines is approximately proportional to the weight of the wool in its various forms

Fig. 209 Yorkshire wool textile towns

The wool textile industry is very important in the northern part of the coalfield. It first developed as a domestic industry using primitive manually-driven 'machines'.

209a What local factors favoured the original development of a wool textile industry in W. Yorkshire? (Think of raw material; water for washing purposes; a form of power.)
209b During the Industrial Revolution (1750–1850) the industry became fully mechanised, and housed in factories around which large towns developed.
  (i) Name three textile machines invented then.
  (ii) What fuel was used to power the steam engines which replaced the crude waterwheels?
  (iii) What advantages did the north have over other traditional areas which also wanted to develop their textile industries?
  (iv) Name the Yorkshire towns involved in the wool textile industry.

Fig. 210 Stages in the production of wool textiles

210a What are the main sources of raw wool today, and what seaports are involved in this trade? (See Fig. 209.)
210b Distinguish the four main stages of wool textile production.
210c Is the wool textile industry expanding or contracting? (Give figures.)

Some limited geographical specialisation occurs in the West Yorks wool textile industry: **Bradford** (main wool market) specialises in worsteds (fine cloths); **Huddersfield** in woollens (rougher cloth); **Halifax** in carpets; and **Batley** and **Dewsbury** in 'shoddy' (a cloth made from waste material). **Leeds** is the commercial centre of West Yorkshire with a variety of industry (e.g. clothing, engineering, leather tanning, brewing, etc.).

210d Suggest some industries closely linked with the wool textile industry. (Think of (i) what the industry requires, and (ii) what happens to the cloth, etc., after it leaves the mills.)

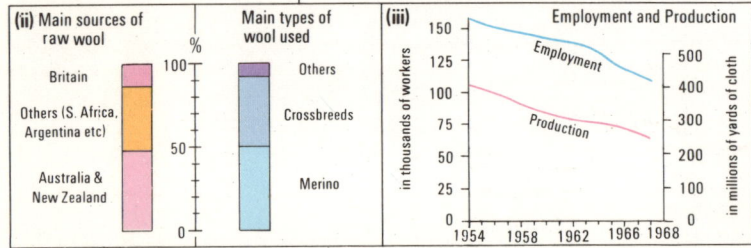

ENGLAND   The Don Valley

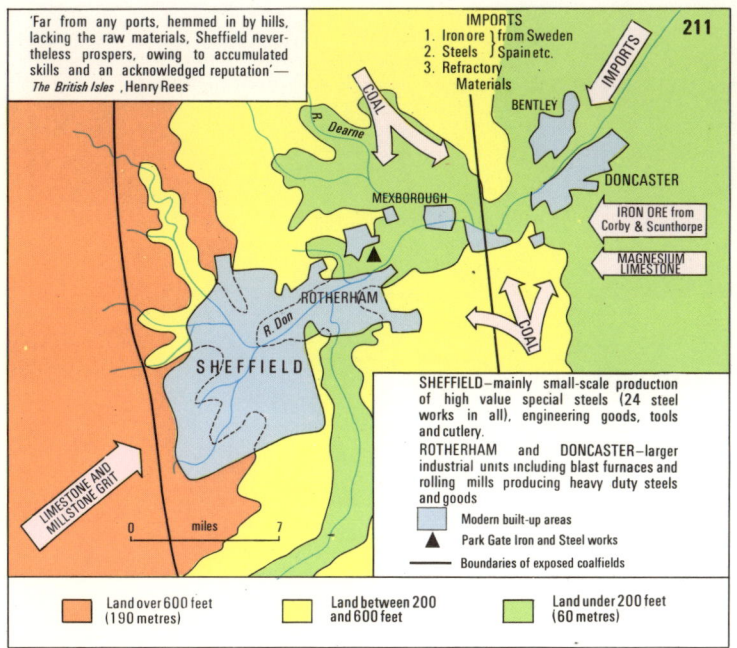

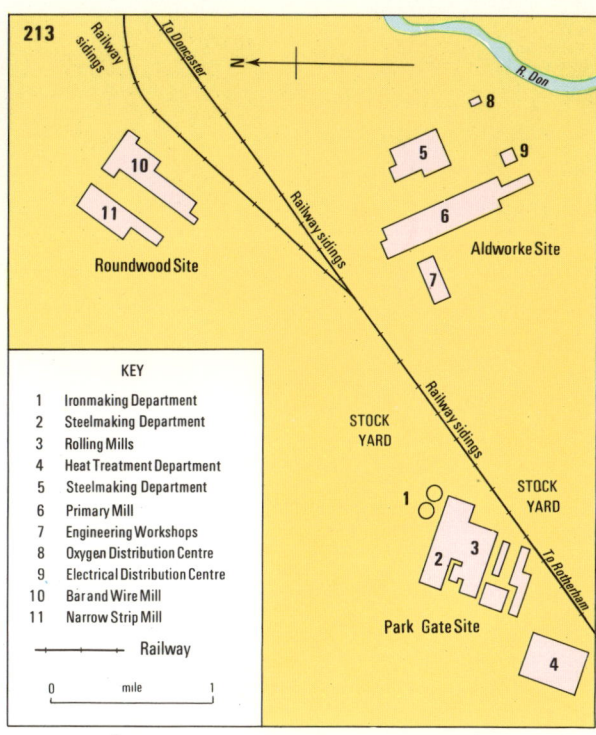

Fig. 211 Iron and steel industry

211a   Where does the river Don enter the sea? (See Fig. 206.)
211b   Name the three largest towns in the Don Valley.
211c   (i) How do the steel industries of Sheffield differ from those in Doncaster and Rotherham?
       (ii) In what way does valley size explain this difference?
211d   (i) What raw materials are used in the iron and steel industry today (think of the crude ore, fuel, flux, refractory materials e.g. furnace linings)?
       (ii) Where are they obtained?
211e   Industrial inertia describes an industry which continues to exist after its original advantages have disappeared. Why can this term be applied to Sheffield's steel industry?

Fig. 212 Development of Sheffield iron and steel industry

212a   (i) What was the original local source of iron ore?
       (ii) Was this local ore of good quality?
       (iii) Suggest why high-grade iron ore and steel were imported as well.
212b   (i) What fuel was used before coal (in the form of coke)?
       (ii) Where was this fuel obtained?
212c   What other raw materials existed locally?
212d   (i) What form of power drove the grindstones, bellows and hammers which were part of the early iron and steel industry?
       (ii) Why was Sheffield well-situated in this respect?

Fig. 213 Park Gate Iron and Steel Works, Rotherham

213a   Park Gate is the only integrated iron and steel works in the Don Valley. Describe its position (see Fig. 212).
213b   What is an integrated iron and steel works?
213c   (i) What raw materials probably occupy the stock yards?
       (ii) Suggest how these raw materials are transported.
213d   Name two other items important in the modern making of iron and steel (one a common gas and the other a form of power).
213e   Name three forms of steel produced at Park Gate.

Fig. 215 Population: employment and growth

215a   (i) What is the approximate population of Sheffield today?
       (ii) During what period was its growth fastest?

Chesterfield, a town to the south of, but closely linked to, the Don valley is another important iron and steel centre.

215b   Scunthorpe is another important iron and steel town. Locate it.
       (i) How does its present population compare with that of Sheffield
       (ii) During what period was Scunthorpe's growth fastest?

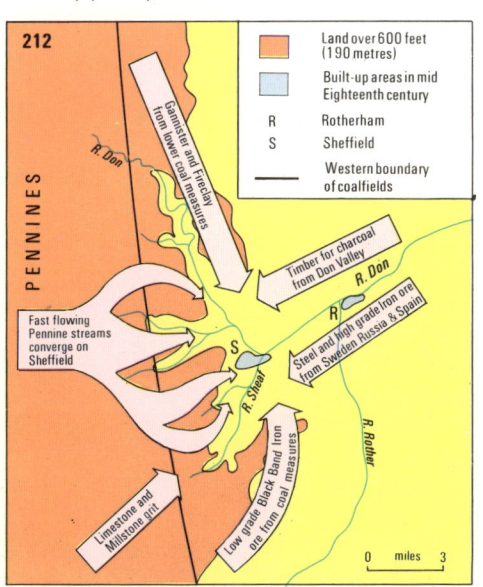

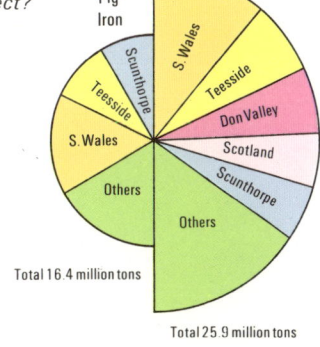

Fig. 214 British iron and steel production, recent year

214a   How does the Don Valley compare with other important crude steel producing regions of Britain?
214b   Suggest one reason why the Don Valley is not independently represented in the semi-circle illustrating British pig-iron production.

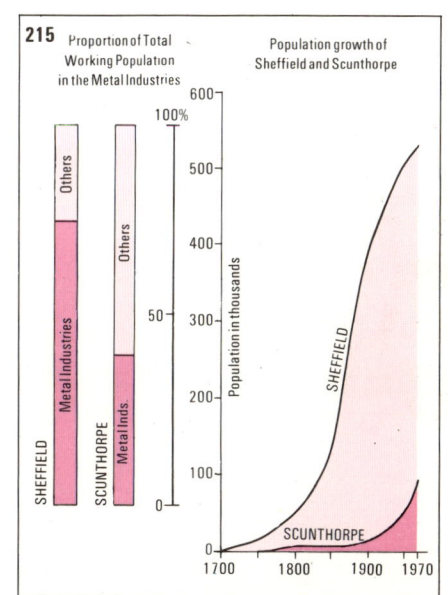

36 ENGLAND Scunthorpe, Humber towns, Nottingham, Derby

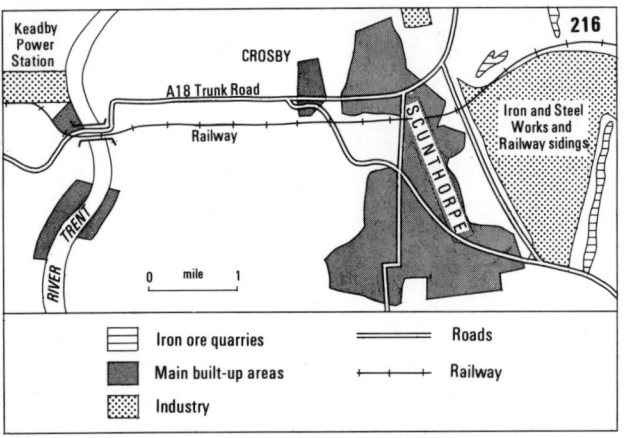

Fig. 216 Scunthorpe: iron and steel centre

216a How far east of the river Trent is Scunthorpe?
216b In what parts of Scunthorpe is the iron and steel industry concentrated?

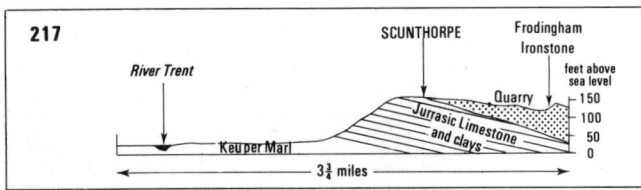

Fig. 217 Section across the Scunthorpe district

217a Some iron and steel centres are situated on coalfields; others are on ironfields.
  (i) Into which category does Scunthorpe fall?
  (ii) What does this suggest about the quality of the local iron ore?
217b With what other rocks does the iron ore occur?

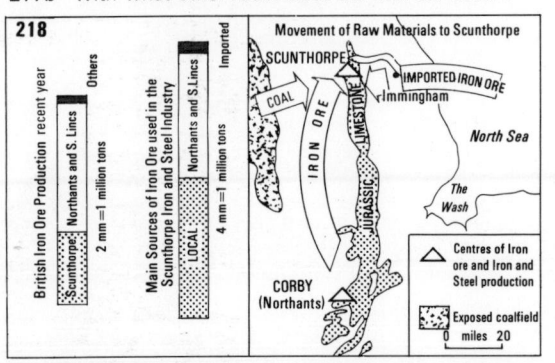

Fig. 218 Sources of Scunthorpe's raw materials

218a How important is the Scunthorpe area as a British producer of (i) iron ore, and (ii) iron and steel? (See also Fig. 214.)
218b Where are the main raw materials for Scunthorpe's iron and steel industry obtained?

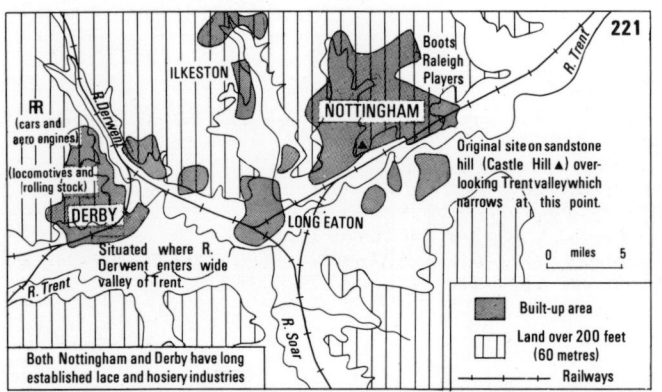

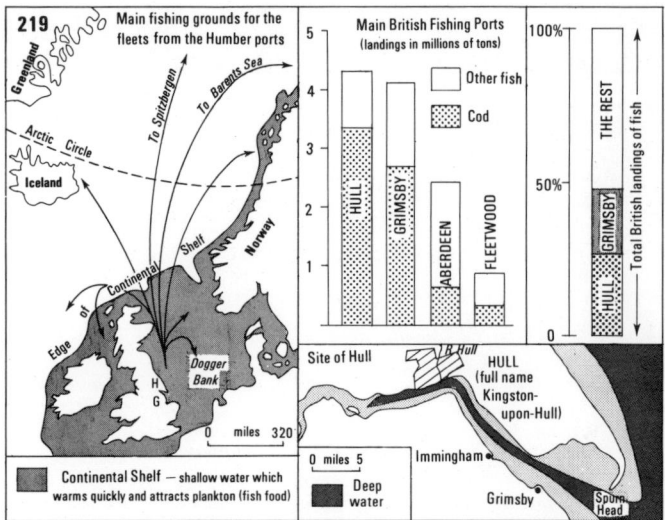

Fig. 219 The Humber: fishing industry (pre Cod War)

219a What percentage of total British landings of fish arrive at (i) Hull, and (ii) Grimsby?
219b How do Hull and Grimsby compare with other fishing ports in Britain?
219c Name some of the main fishing grounds visited by trawlers from the Humber ports.
219d What is the significance of the Continental Shelf for the fishing industry?
219e Which part of it lies nearest to the Humber?
219f Suggest three industries which make equipment for the fishing industry.
219g (i) What was Hull's former name?
  (ii) Distinguish one reason why Hull has become a much larger town than either Grimsby or Immingham.

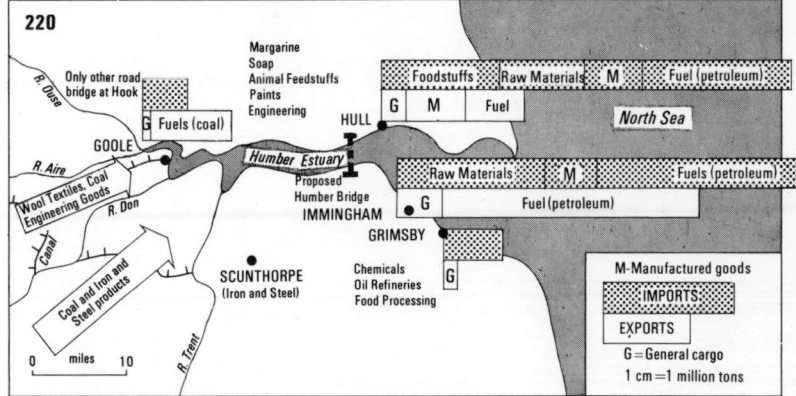

Fig. 220 The Humber: port traffic and industry

220a The Humber towns are also important seaports. Which one handles most trade?
220b In what two ways does the cargo traffic of Hull differ from that of Immingham?
220c The hinterland of a port is the area served by that port. Outline the hinterland of the Humber ports.
220d The area has
  (i) good facilities for importing raw materials and exporting manufactured goods,
  (ii) large areas of flat undeveloped land.
Name some of the industries in the area and indicate which are based on imported raw materials.

Fig. 221 Nottingham and Derby

221a Account for the site and situation of (i) Nottingham, (ii) Derby.
221b What industry has long been important in both Nottingham and Derby?
221c Name (i) five well-known firms in this area,
  (ii) their products,
  (iii) where they are located.
221d Why is this area a route centre?

NORTHERN ENGLAND Agriculture

Northern England—Agriculture

Fig. 222 The counties of Northern England

222a Draw two columns—one headed 'west', the other headed 'east', and list the counties of Northern England accordingly.

Fig. 223 Agricultural land use

223a Where are most cereals grown in Northern England—east or west of the Pennines?

223b What do the words 'arable farming' mean?

223c Which is the only 'county' in Northern England with over half its agricultural land devoted to arable farming?

223d What county has a large proportion of temporary and permanent grassland but a relatively small proportion of highland?

223e Is there any apparent relationship between the proportion of rough grazing land and the proportion of highland in a county? Quote two examples.

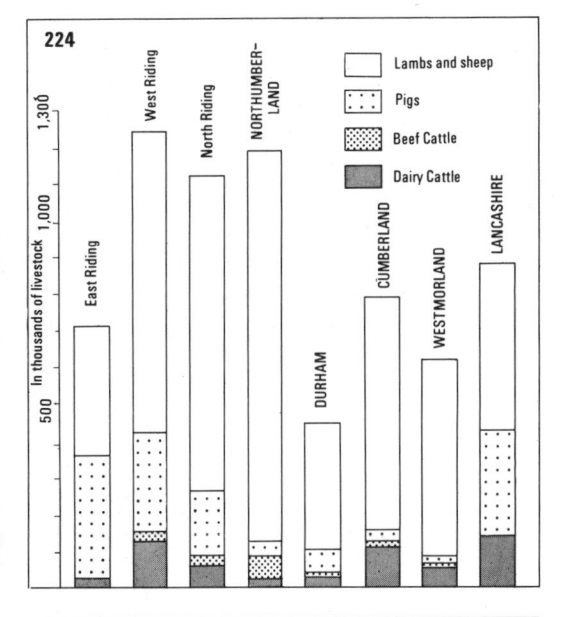

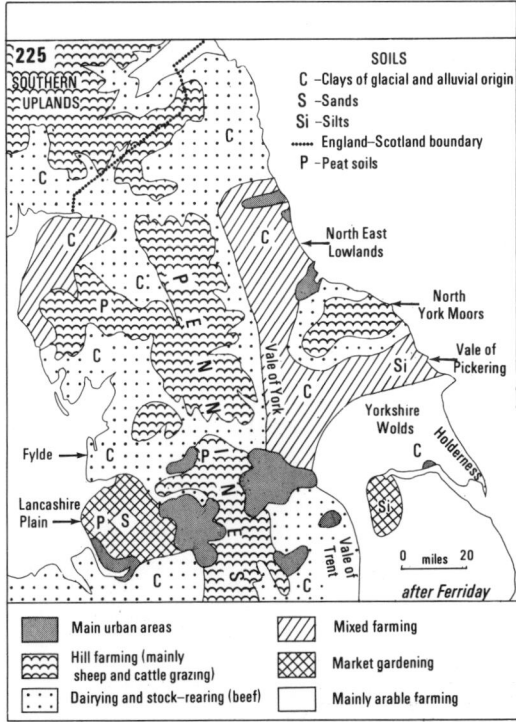

Fig. 224 Livestock

224a What is the most common kind of livestock in Northern England?

224b (i) What are the three main sheep-rearing counties in Northern England?
(ii) Is there any apparent link between a large number of sheep and the proportion of highland in a county? Quote an example (see Fig. 223).

224c What is the main beef-cattle rearing county in Northern England?

224d (i) Name the three main dairying counties.
(ii) Is dairying more important east or west of the Pennines?

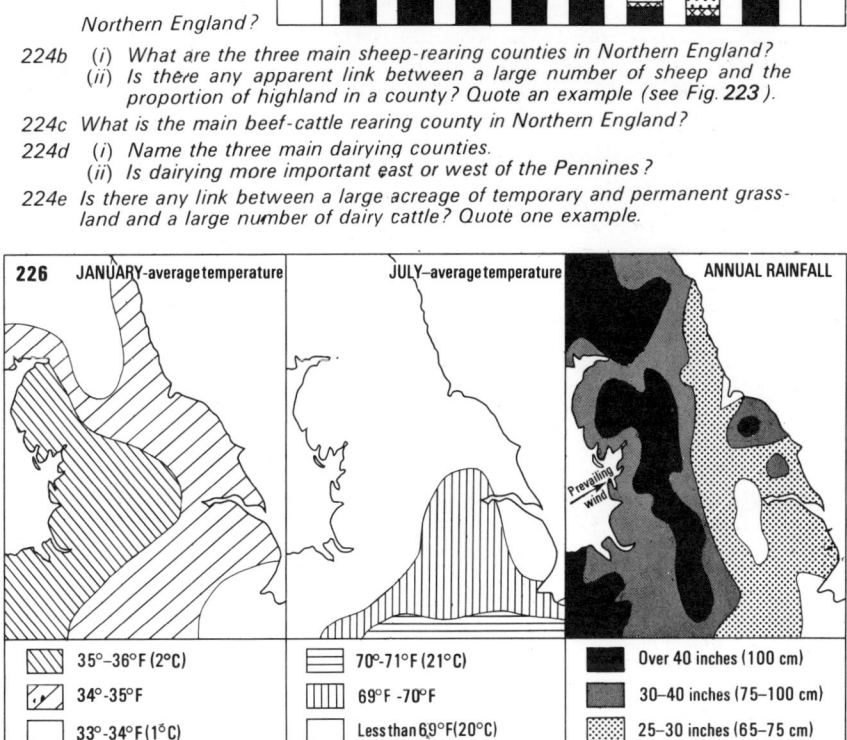

224e Is there any link between a large acreage of temporary and permanent grassland and a large number of dairy cattle? Quote one example.

Figs. 225/226 Agricultural regions and climate

225/226a Lancashire is the most important milk-producing county in Northern England.
(i) Name Lancashire's main dairying areas.
(ii) Indicate the climatic and soil conditions that exist in these areas.

225/226b What towns are supplied with milk from
(i) The Fylde, and   (ii) the Vale of Trent?

225/226c Describe the types of farming shown here east of the Pennines and the climates and soils in each

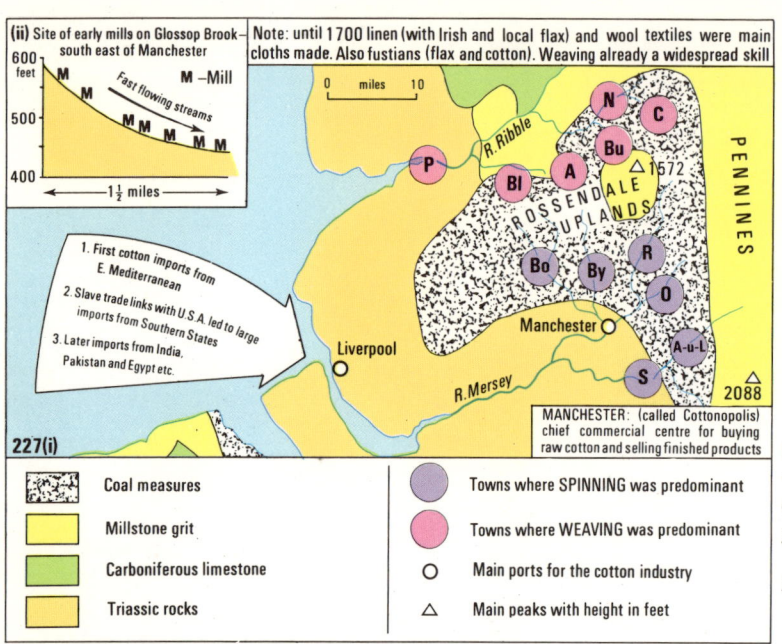

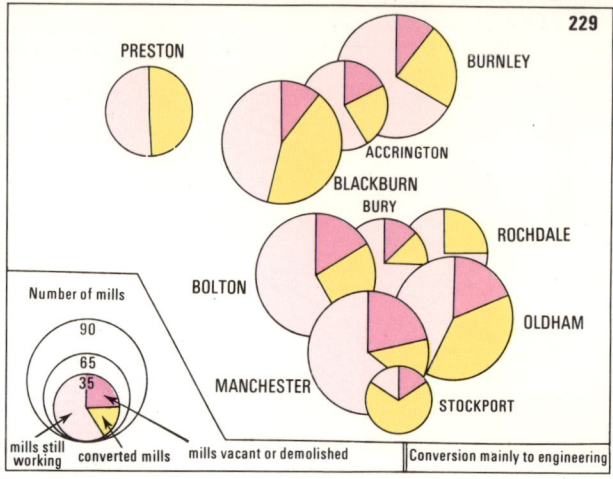

ENGLAND S. Lancs./Merseyside / Greater Manchester / Cheshire Industrial Region

Fig. 227 Development of the 'Lancashire' cotton industry

227a What original factors favoured the development of a cotton textile industry in East Lancashire? (Think of textile traditions, suitable water for washing purposes, power, etc.)

227b (i) Where was the raw cotton first obtained?
(ii) Under what circumstances did the import of American cotton arise?
(iii) What advantage had this area for importing cotton?

227c The mechanisation of the cotton textile industry during the Industrial Revolution was eventually accompanied by a change to steam power. What local fuel assisted this change?

227d A limited specialisation occurred in the Lancashire cotton textile industry.
(i) Using two columns, distinguish the 'weaving' and 'spinning' towns.
(ii) Name the upland that separates them.

227e (i) What function has Manchester in the cotton textile industry?
(ii) What advantages did Manchester have as a route centre?
(iii) What equivalent has Manchester in the Yorkshire wool textile industry? (See Fig. 209, p. 34.)

Fig. 228 British cotton exports, 1910–70

228a What were the main foreign markets for British cotton textiles?

228b Explain Britain's export performance in cotton textiles to India during the twentieth century.

The decline in cotton textile exports has been due to competition from
(i) former markets, e.g. India,
(ii) the Japanese textile industry,
(iii) man-made fibres.

Fig. 229 Decline of the 'Lancashire' cotton industry

The Cotton Industry Act of 1959 led to an organised running down of the cotton textile industry.

229a In which four Lancashire towns has the cotton textile industry declined most?

229b Demolition of disused mills has taken place, but conversion has been most common.
(i) In what two towns have most conversions occurred?
(ii) What industry often replaces the cotton industry?

The area specialises today in high quality cotton goods, which are able to compete more easily in world markets.

Fig. 230 Main industries

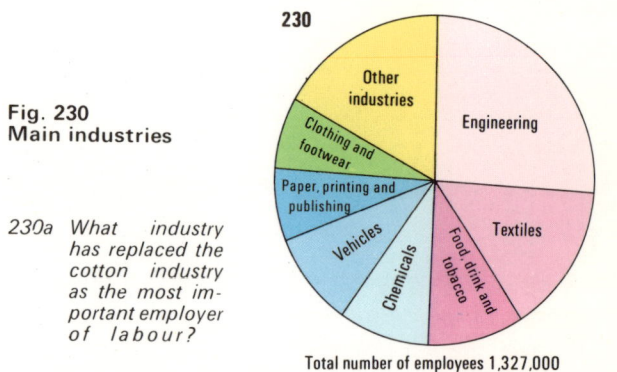

230a What industry has replaced the cotton industry as the most important employer of labour?

230b Approximately how many people work in the engineering industry?

230c Name three other important industries in this region.

Fig. 231 Coalmining

231a In what two parts of the Lancashire coalfield, and near which towns, are today's collieries found?

231b What proportion of total British coal production comes from this field?

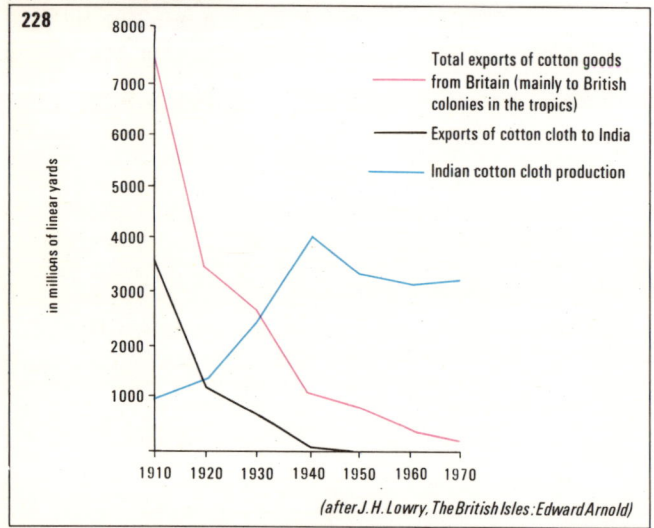

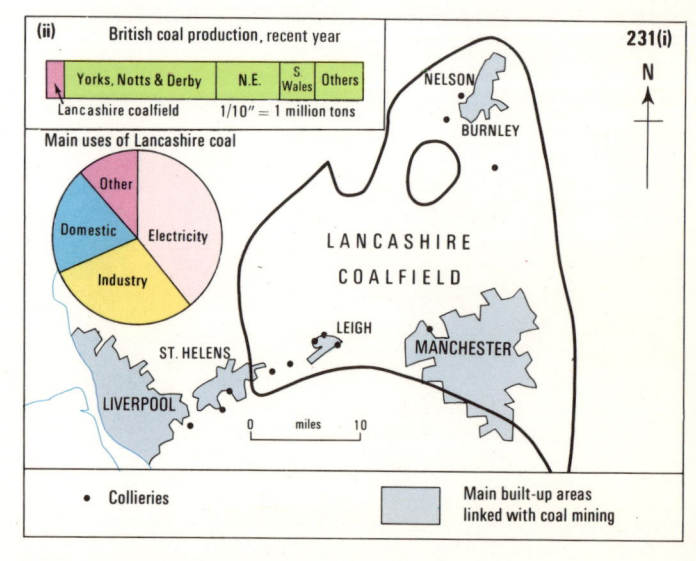

ENGLAND S.Lancs./Merseyside / Greater Manchester / Cheshire Industrial Region

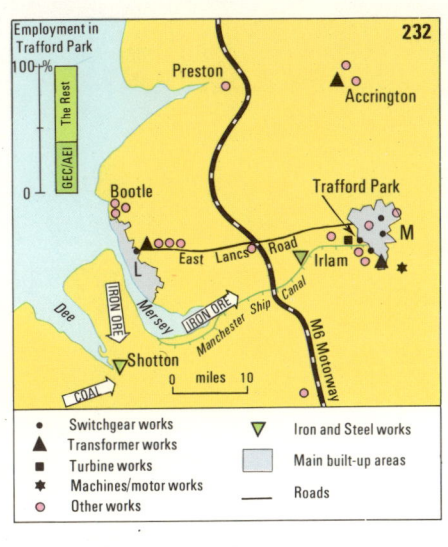

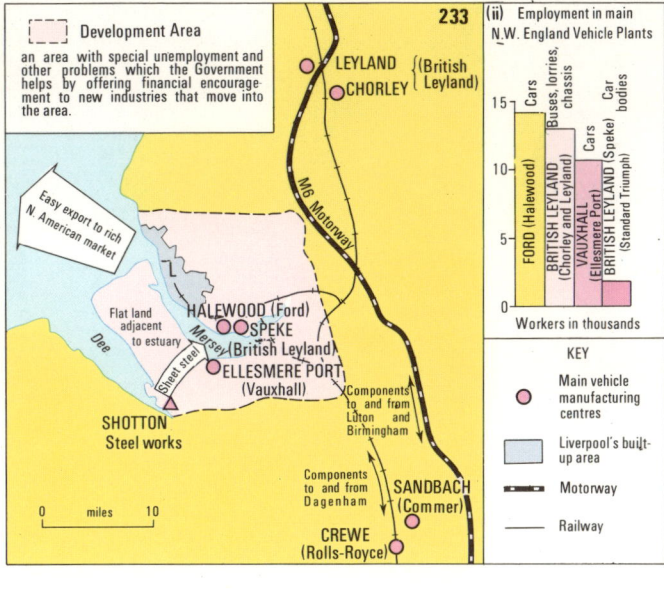

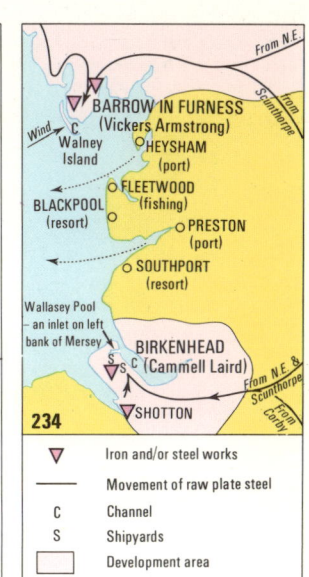

Fig. 232 Locations of the G.E.C./E.E./A.E.I Group works

232a The engineering industry grew up in response to demand from newly-mechanised industries during the Industrial Revolution. What sort of equipment would have been supplied to (i) the coal industry, (ii) the textile industry, (iii) the shipbuilding industry, (iv) the new railway companies?

232b Some of these engineering industries have declined. Suggest why.

232c The electrical engineering industry is today the most rapidly expanding industry in the north-west. The General Electric/English Electric/Associated Electrical Industries group dominates this industry.
 (i) Name the three areas on fig 232 where this group has most of its works.
 (ii) Name three electrical engineering goods produced.

Some other important electrical engineering firms in the area include Lucas (Burnley), Plessey (Wigan), Ferranti (Oldham and Manchester) and Hawker-Siddeley (Manchester).

Figs. 235 and 236 Cheshire and Merseyside chemical industries

235a What is Cheshire's most valuable raw material and how is it extracted?

235b What towns are engaged in the Cheshire salt industry?

235c The chemical industry receives most of the salt.
 (i) What are the three main centres of this industry?
 (ii) Which one is located on the saltfield?

235d (i) What do the initials 'I.C.I.' stand for?
 (ii) Outside what town is I.C.I.'s main Weston Point plant?

235e (i) What other raw materials are used in the chemical industry?
 (ii) Where are they obtained?

Typical products of the chemical industry include caustic soda, sulphuric acid, bleaches, dyes, etc. Many of these products are used in related industries, e.g. soap- and detergent-making, leather tanning, glass-making and the rubber and paper industries.

235f (i) What oil company has an important refinery at Stanlow?
 (ii) Where is most of the petroleum obtained?

235 DIAGRAMMATIC SECTION OF A CHESHIRE SALT BED

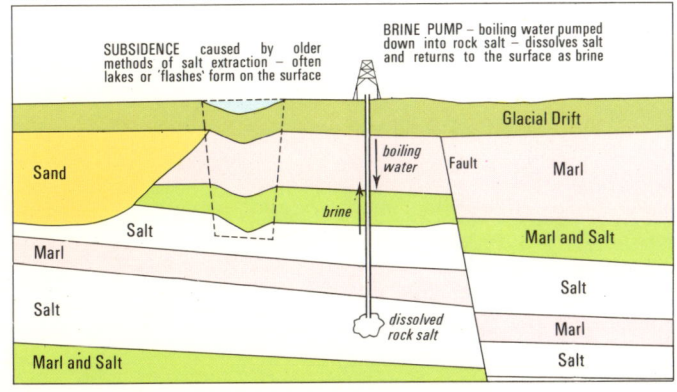

Fig. 233 The vehicle-assembly industry

233a How does this industry rank in North-West England? (See Fig. 230, p. 38.)

233b (i) What firms have vehicle-assembly or component plants in North-West England?
 (ii) Name the two largest plants and the number of workers at each.

233c What advantages favour the establishment of a vehicle-assembly industry on Merseyside? (Think of government aid, labour, communications, suitable sites, export facilities, 'raw' materials and components.)

All the vehicle-assembly plants on Merseyside are post-1950 developments.

Fig. 234 The shipbuilding industry

234a (i) In what two towns is the shipbuilding industry concentrated?
 (ii) Name the firms at each.

Repairing, as well as building, is important today. Polaris nuclear-powered submarines are an example of major launchings from these yards in recent years.

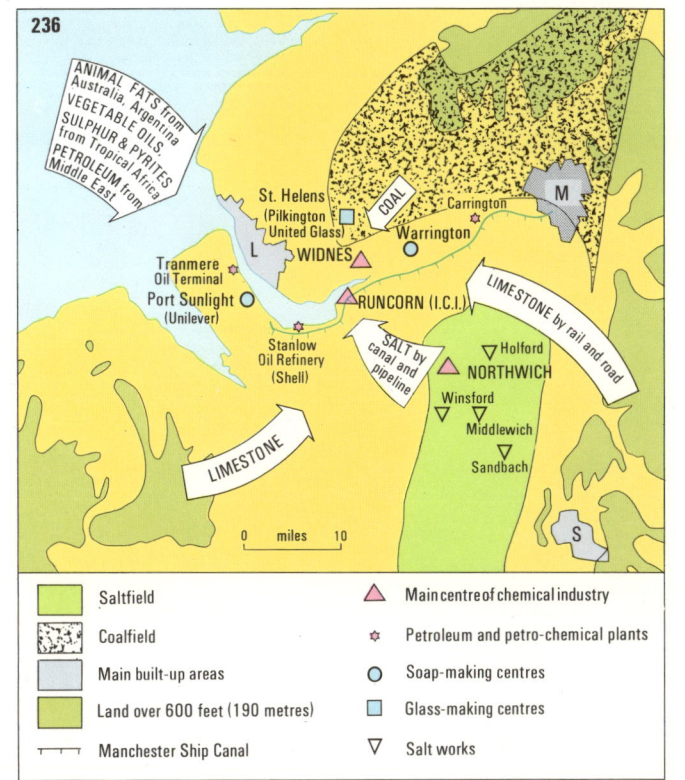

ENGLAND Manchester and Liverpool

237(i) POPULATION OF MAIN NORTHERN TOWNS
(Liverpool, Manchester, Leeds, Sheffield, Newcastle, Bradford)

(ii) NEW TOWNS IN LANCASHIRE & CHESHIRE
Leyland, Chorley, Skelmersdale, Kirkby, Risley, Runcorn

Fig. 237 Population

237a What are the two largest cities in Northern England?

237b If the population of London is 8·5 million, how many times bigger is it than Liverpool?

237c How and where are Liverpool and Manchester trying to solve their housing problems?

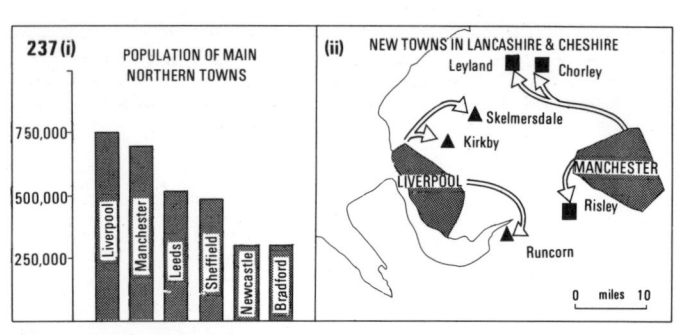

Figs. 238 and 239 The early development of Liverpool

238/239a (i) What was the main port in North-West England until 1700?
(ii) Why did it decline?

238/239b Why did the Mersey estuary remain free of silt?

238/239c What stimulated Liverpool's first phase of growth?

238/239d Draw a diagram including the missing side of the Slave Trade Triangle.

238 LIVERPOOL IN 17th CENTURY
Early trade with Ireland and Scotland. 'bottle-neck' shape of estuary > powerful tidal action – prevented silting up. LIVERPOOL Grew up on sandstone ridge above an inlet called 'The Pool'. Wide, open estuary – weak tidal action > silting up until eventually Chester isolated from open sea. CHESTER Premier port in N.W. England until 1700 – silting up.

239 18th CENTURY DEVELOPMENT OF LIVERPOOL
Sugar, molasses and rum from WEST INDIES in exchange for slaves. Trinkets, alcohol and guns etc. to WEST AFRICA in exchange for slaves. LIVERPOOL 1st phase of rapid growth based on 'SLAVE TRADE TRIANGLE' – foundation of Nineteenth century Transatlantic trade links. CHESTER – now isolated from the open sea.

Fig. 241 Dock basins along the Mersey

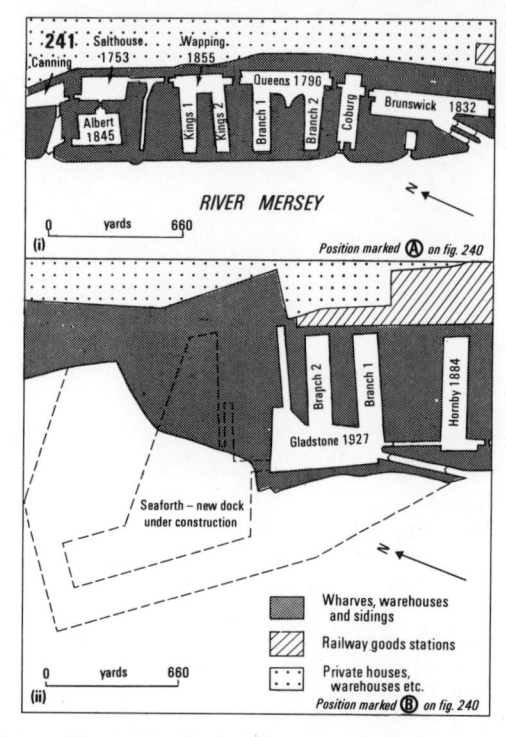

(i) Salthouse 1753, Canning, Wapping 1855, Albert 1845, Kings 1, Kings 2, Queens 1790, Branch 1, Branch 2, Coburg, Brunswick 1832. RIVER MERSEY. Position marked A on fig. 240.

(ii) Branch 2, Branch 1, Hornby 1884, Gladstone 1927, Seaforth – new dock under construction. Wharves, warehouses and sidings; Railway goods stations; Private houses, warehouses etc. Position marked B on fig. 240.

241a Study (i) the positions (Fig. 240), (ii) the dimensions, and (iii) the dates of construction of the dock basins and explain in what way these three factors are related.

Fig. 242 British port traffic

242a How many times is the traffic of the port of London greater than that of
(i) Liverpool?
(ii) Manchester?

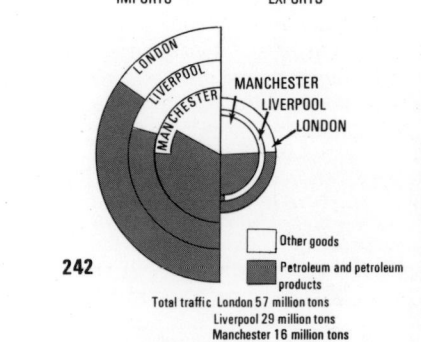

242 IMPORTS / EXPORTS (London, Liverpool, Manchester). Other goods; Petroleum and petroleum products.
Total traffic London 57 million tons, Liverpool 29 million tons, Manchester 16 million tons.

Fig. 243 Main non-fuel traffic of ports of Liverpool and Manchester

243a Distinguish two raw materials and two foodstuffs imported through Liverpool and Manchester.

243b In which of the following categories would you place the main exports from these ports: raw materials, foodstuffs, fuels, manufactured goods?

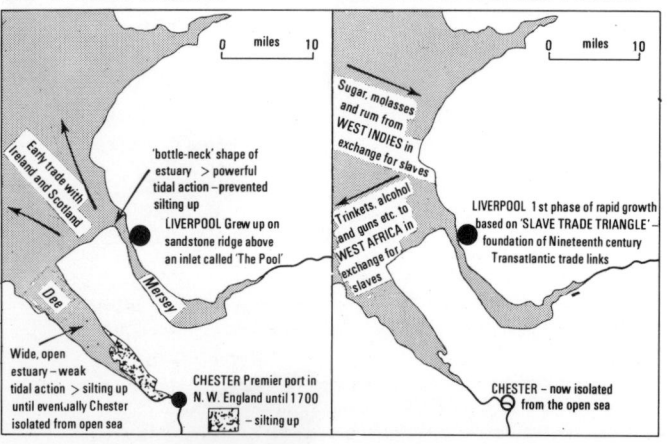

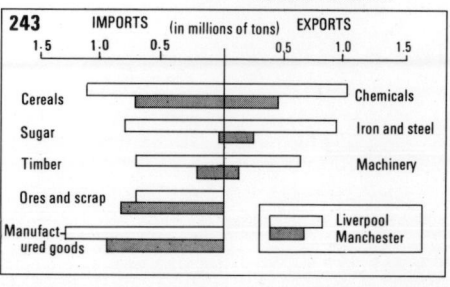

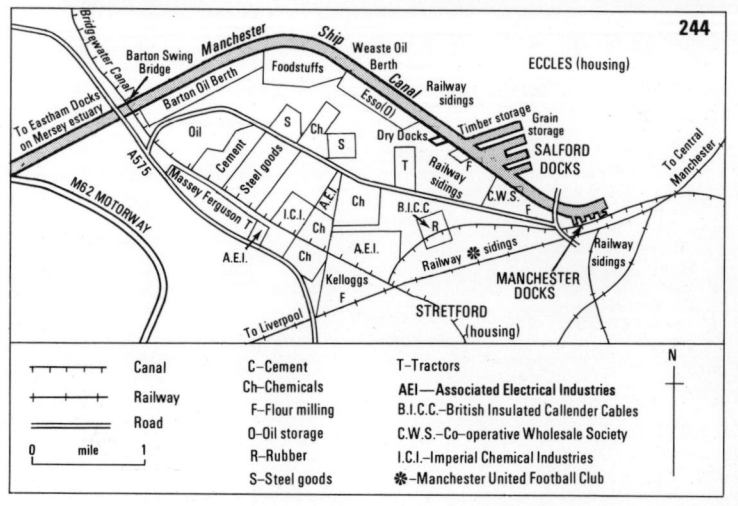

243 IMPORTS (in millions of tons) EXPORTS
Cereals — Chemicals
Sugar — Iron and steel
Timber — Machinery
Ores and scrap
Manufactured goods
Liverpool / Manchester

Fig. 240 The 19th century development of Liverpool and Manchester

240a What trade stimulated the main period of dock construction in Liverpool, and when?

240b What is the significance of the year 1894 in Manchester's history?

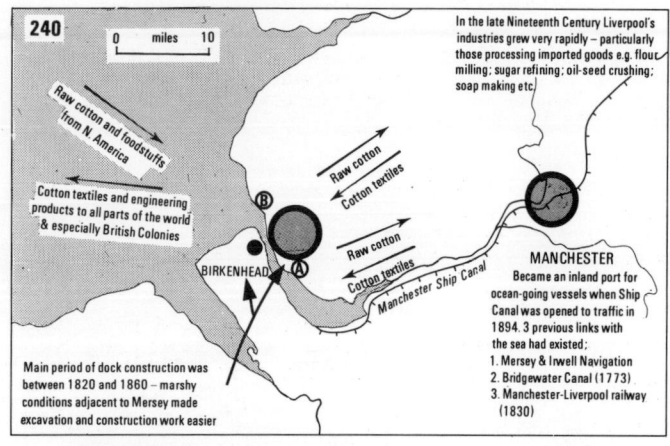

240 Raw cotton and foodstuffs from N. America. Cotton textiles and engineering products to all parts of the world & especially British Colonies. BIRKENHEAD. Raw cotton, Cotton textiles. Manchester Ship Canal. MANCHESTER Became an inland port for ocean-going vessels when Ship Canal was opened to traffic in 1894. 3 previous links with the sea had existed:
1. Mersey & Irwell Navigation
2. Bridgewater Canal (1773)
3. Manchester-Liverpool railway (1830)

Main period of dock construction was between 1820 and 1860 – marshy conditions adjacent to Mersey made excavation and construction work easier.

In the late Nineteenth Century Liverpool's industries grew very rapidly – particularly those processing imported goods e.g. flour milling; sugar refining; oil-seed crushing; soap making etc.

Manchester has many important functions. It is an important route centre and has consequently developed commercial, shopping and banking interests. It is also an inland port and industrial centre.

Fig. 244 Trafford Park

244a Trafford Park is one of the most important industrial areas in Manchester. Name:
(i) three well-known firms with factories there.
(ii) the type of product each produces.

244 Bridgewater Canal, Barton Swing Bridge, Manchester Ship Canal, Weaste Oil Berth, Barton Oil Berth, Foodstuffs, Esso(O), Railway sidings, Timber storage, Grain storage, SALFORD DOCKS, To Eastham Docks on Mersey estuary, Oil, Cement, Massey Ferguson, Steel goods, Dry Docks, A575, M62 MOTORWAY, I.C.I., Ch, A.E.I., B.I.C.C., R, C.W.S., To Central Manchester, Kelloggs, F, A.E.I., Railway sidings, MANCHESTER DOCKS, STRETFORD (housing), To Liverpool, ECCLES (housing).

Canal; Railway; Road
C–Cement; Ch–Chemicals; F–Flour milling; O–Oil storage; R–Rubber; S–Steel goods; T–Tractors
AEI—Associated Electrical Industries
B.I.C.C.—British Insulated Callender Cables
C.W.S.—Co-operative Wholesale Society
I.C.I.—Imperial Chemical Industries
✳—Manchester United Football Club

ENGLAND Industrial North-East

COAL-MINING; IRON AND STEEL; SHIPBUILDING

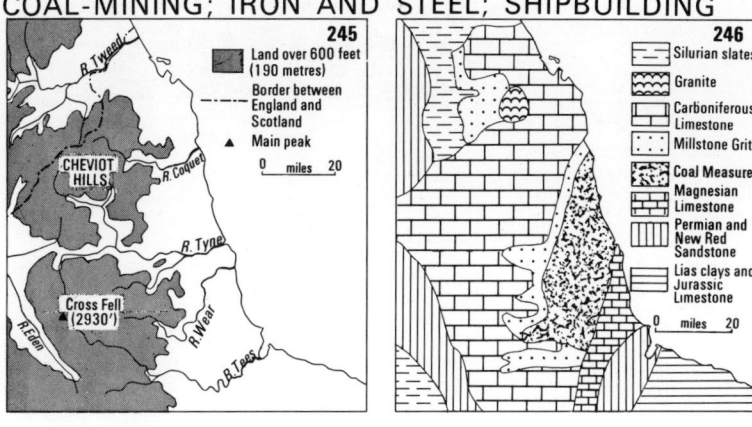

Fig. 245 Relief

245a What river has produced the main gap through the Pennines in Northern England?

245b What is the width (in miles) of the coastal plain in the north (River Coquet) and south (River Tees) of this region?

Fig. 246 Geology

246a Which rock in North-East England is particularly valuable to man?

246b Compare the rocks the rivers Tyne and Tees flow over (from source to mouth).

Fig. 247 Northumberland and Durham coalfield

247a Describe the boundaries and calculate the dimensions of the Northumberland and Durham coalfield.

247b Make a simple copy of the cross-section and indicate the exposed and concealed parts of this coalfield.

247c Where are the main collieries located today?

247d What are the main two uses of coal?

247e Indicate the destinations of coal that is 'exported', and four ports concerned with this traffic.

247f (i) What is the annual coal production of this field?
(ii) What proportion of total British coal production does it represent?

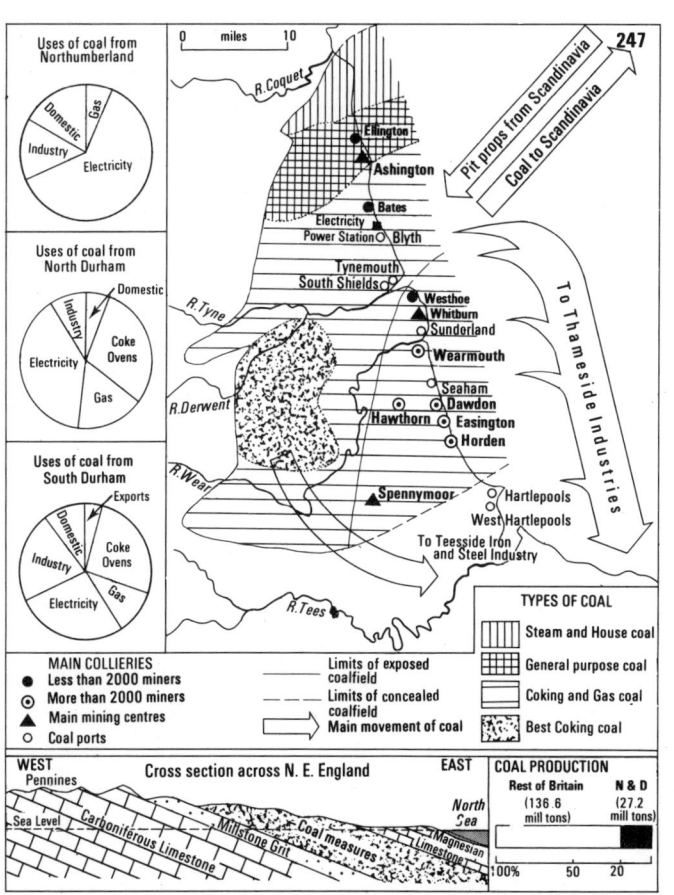

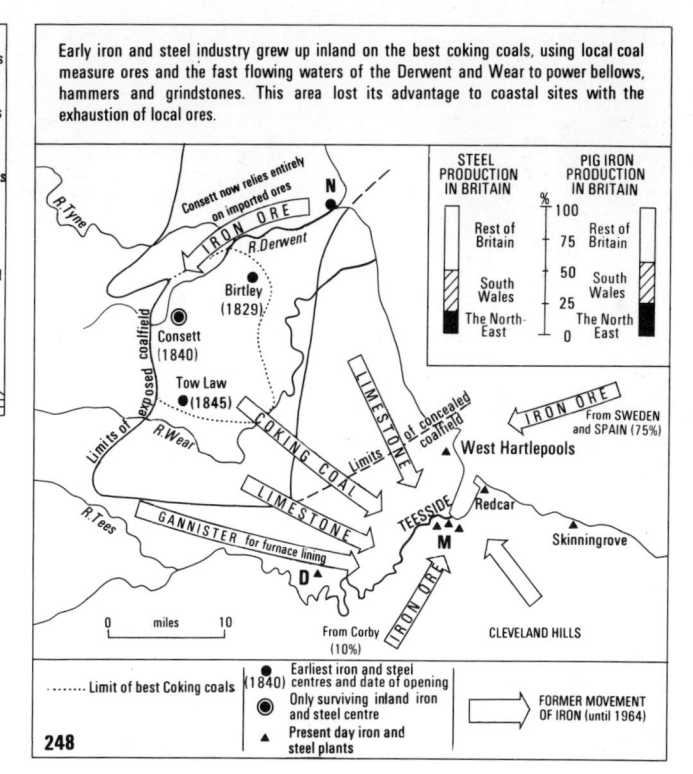

Early iron and steel industry grew up inland on the best coking coals, using local coal measure ores and the fast flowing waters of the Derwent and Wear to power bellows, hammers and grindstones. This area lost its advantage to coastal sites with the exhaustion of local ores.

Fig. 248 The iron and steel industry

248a List three reasons why the iron industry first grew up inland.

248b Which of the earliest centres still produces iron and steel?

248c Name four other important modern iron and steel towns.

248d Explain why the new iron and steel works are all on the coast. (Think in terms of supply of raw material, suitable sites, etc.)

248e Where are the 'other' raw materials for these works obtained?

248f Is the North-East a more important steel or pig-iron producing area?

248g How does iron and steel production in the North-East compare with that of (i) South Wales, (ii) the rest of Britain?

Fig. 249 Engineering

249a Suggest one of the main raw materials used in the heavy engineering industry.

249b (i) Name three engineering centres in North-East England.
(ii) Give examples of what they produce.

The thirty-six industrial estates in North-East England have been established in response to a high unemployment rate resulting from the decline of coal-mining, the iron and steel industry and shipbuilding. These estates are centres of light industrial development.

Fig. 250 Shipping output

250a What is the national importance of North-East England as a producer of merchant ships?

250b (i) What are the three main rivers concerned?
(ii) How has output changed at each since 1958?

The main type of vessel produced today is the giant oil tanker.

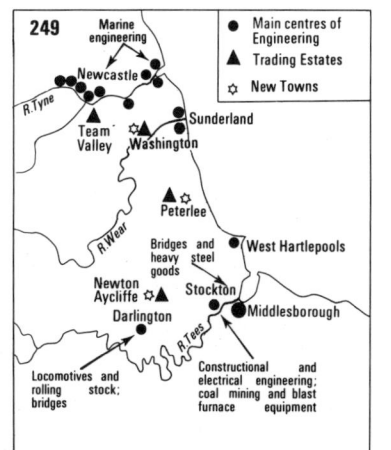

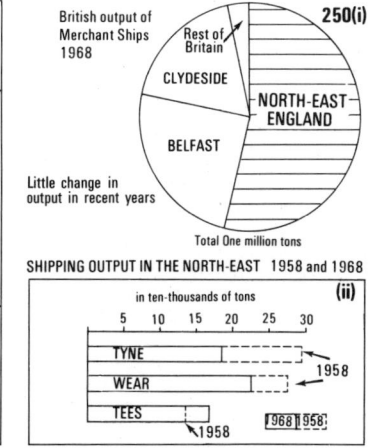

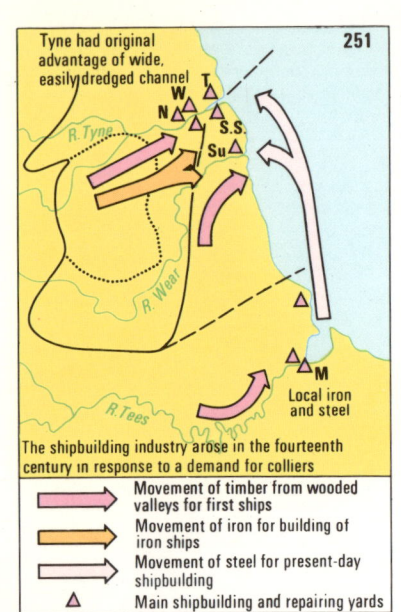

ENGLAND Tyneside and Teesside

Fig. 251 The development of the shipbuilding industry

Map shows: Tyne had original advantage of wide, easily dredged channel. R. Tyne with points W, T, N, S.S., Su. R. Wear. R. Tees with M. Local iron and steel.

The shipbuilding industry arose in the fourteenth century in response to a demand for colliers.

Legend:
- Movement of timber from wooded valleys for first ships
- Movement of iron for building of iron ships
- Movement of steel for present-day shipbuilding
- Main shipbuilding and repairing yards

251a What originally stimulated the development of a shipbuilding industry in the North-East?

251b (i) Describe the succession of basic materials used in this industry since the 14th century.
(ii) Where were they obtained?

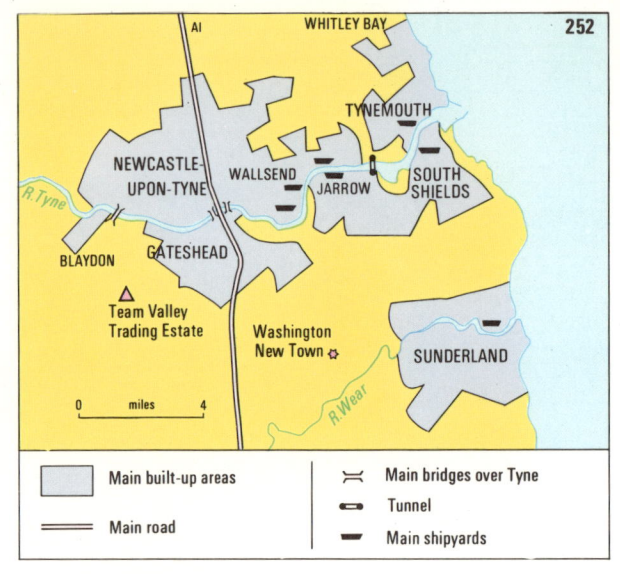

Fig. 252 Tyneside and Sunderland

Map shows: A1, Whitley Bay, Newcastle-upon-Tyne, Wallsend, Tynemouth, Jarrow, South Shields, Blaydon, Gateshead, Team Valley Trading Estate, Washington New Town, Sunderland, R. Tyne, R. Wear.

Legend:
- Main built-up areas
- Main road
- Main bridges over Tyne
- Tunnel
- Main shipyards

252a Explain why there are more shipyards on the Tyne than on the Wear (see also Fig. 251).

252b Suggest two reasons why a tunnel rather than a bridge is the only crossing point of the Tyne below Newcastle.

Newcastle is the regional capital of North-East England, and the natural focus for the region. It is also an important industrial centre and seaport. The built-up area from Newcastle to South Shields is a conurbation.

252c Explain 'conurbation'.

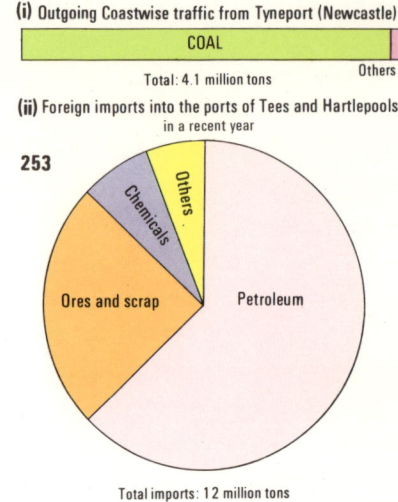

Fig. 253 Tyneside and Teesside imports and exports

(i) Outgoing Coastwise traffic from Tyneport (Newcastle)
COAL — Others
Total: 4.1 million tons

(ii) Foreign imports into the ports of Tees and Hartlepools in a recent year
Pie chart: Petroleum, Ores and scrap, Chemicals, Others
Total imports: 12 million tons

253a (i) What is Newcastle's main coastwise export?
(ii) What is its likely destination? (See Fig. 247, p. 41.)

253b (i) List Teesside's imports in order of importance.
(ii) Indicate which of the following categories they come under: raw material, fuel, foodstuffs, manufactured goods.

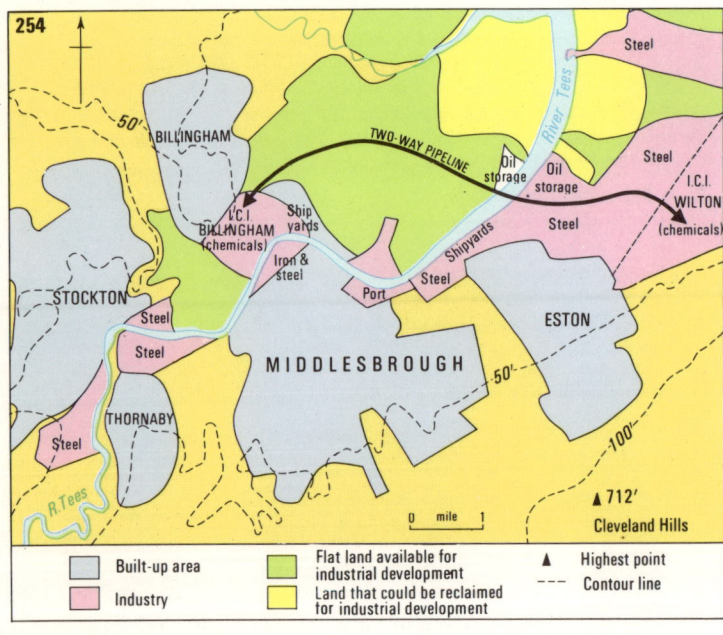

Fig. 254 Teesside: main towns and industries

Map shows: Billingham, I.C.I. Billingham (chemicals), Stockton, Thornaby, Middlesbrough, Eston, I.C.I. Wilton (chemicals), River Tees, Two-way pipeline, Oil storage, Steel, Shipyards, Iron & steel, Port, Cleveland Hills 712'.

Legend:
- Built-up area
- Industry
- Flat land available for industrial development
- Land that could be reclaimed for industrial development
- Highest point
- Contour line

254a What towns make up Teesside?
254b List the three main industries on Teesside.
254c Name and locate (i) the two largest chemical plants and (ii) the company that operates them.
254d Suggest some advantages favouring heavy industrial development on Teesside. (Think in terms of suitable sites, labour, import and export facilities, fuel and water.)

Fig. 255 Teesside chemical industry

255a List the raw materials used in the Teesside chemical industry and indicate whether they occur locally or are imported.

255b Name two industries that use finished products from the I.C.I. chemical plants.

Fig. 256 Teesside population growth

256a What is the present population of Middlesbrough?
256b What percentage of the total population of Teesside live in Middlesbrough?
256c During which years did Middlesbrough grow fastest?
256d What was the main reason for this rapid increase of population?

Fig. 255 Simple flow diagram of raw materials and products of the two I.C.I. chemical plants at Billingham and Wilton

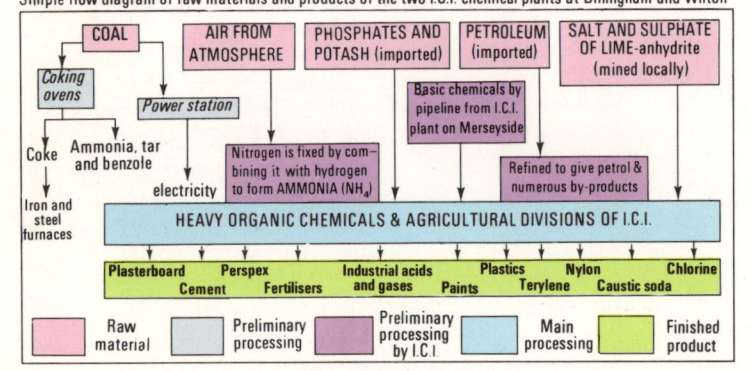

Raw materials:
- COAL → Coking ovens → Coke (to Iron and steel furnaces), Ammonia, tar and benzole; Power station → electricity
- AIR FROM ATMOSPHERE → Nitrogen is fixed by combining it with hydrogen to form AMMONIA (NH₄)
- PHOSPHATES AND POTASH (imported)
- PETROLEUM (imported) → Basic chemicals by pipeline from I.C.I. plant on Merseyside; Refined to give petrol & numerous by-products
- SALT AND SULPHATE OF LIME — anhydrite (mined locally)

→ HEAVY ORGANIC CHEMICALS & AGRICULTURAL DIVISIONS OF I.C.I.

Finished products: Plasterboard, Cement, Perspex, Fertilisers, Industrial acids and gases, Plastics, Paints, Nylon, Terylene, Caustic soda, Chlorine

Legend: Raw material | Preliminary processing | Preliminary processing by I.C.I. | Main processing | Finished product

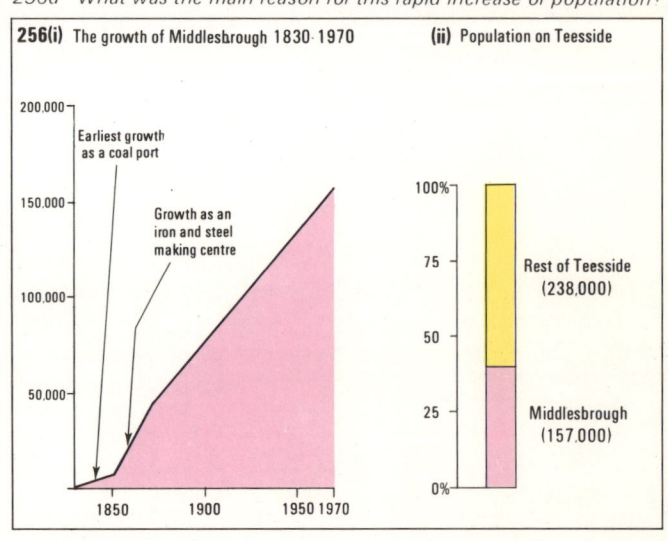

Fig. 256(i) The growth of Middlesbrough 1830–1970

Graph from 0 to 200,000 (1830–1970). Earliest growth as a coal port. Growth as an iron and steel making centre. Reaches ~157,000 by 1970.

(ii) Population on Teesside

Bar chart:
- Rest of Teesside (238,000)
- Middlesbrough (157,000)

WALES General 43

257 Relief

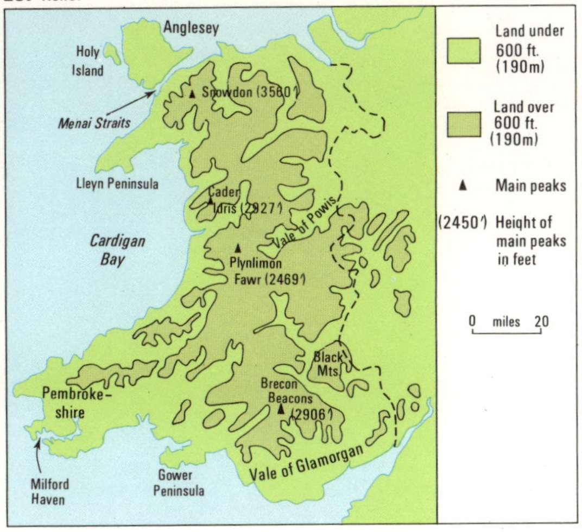

Fig. 257 Relief

257a Name the two estuaries between which the English-Welsh boundary runs.

257b Indicate the name, height and position of the two highest peaks in Wales.

257c (i) Is Wales a predominantly highland or lowland country?
(ii) Indicate the main areas of lowland

Fig. 258 Annual rainfall

Most rainfall in Britain is either relief rainfall or cyclonic rainfall. Convection rainfall mainly occurs in E. England during the summer months.

258a What parts of Wales receive more than eighty inches of rainfall each year?

258b Draw a simple cross-section of North Wales (west-east) and label (i) the direction of the prevailing winds, (ii) the area of highest rainfall, (iii) the rain-shadow area.

258 Rainfall

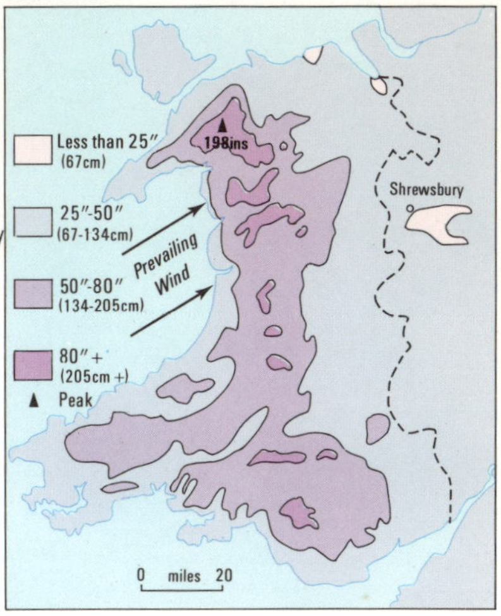

Fig. 259/260 Agricultural land-use: and livestock

259/260a What percentage of total agricultural land in Wales is permanent or temporary pasture?

259/260b Explain the large proportion of rough grazing land and the small proportion of land under cereals in Wales (see Figs. 257 and 258).

259/260c Locate the old Montgomeryshire and Pembrokeshire in an atlas.
(i) Describe the position of each county,
(ii) Indicate which has most lowland.

259/260d Outline the main differences between the agriculture of a highland and lowland Welsh county and suggest why these differences occur.

259

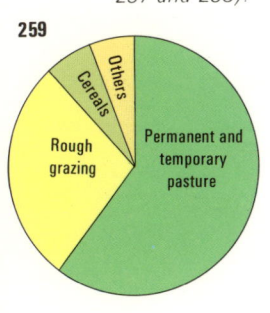

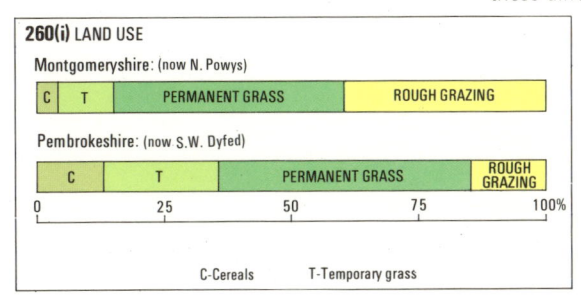

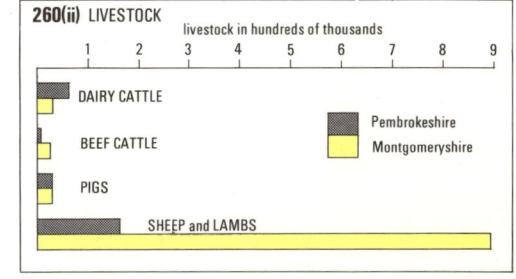

260(i) LAND USE

260(ii) LIVESTOCK

Fig. 261 Main agricultural areas

261a Indicate the main types of farming in Wales, using the following three headings: Highlands, Coastal Plain, Interior Valleys.

261b Note down any exceptions that do not fit in under any of these headings.

Fig. 262 Water supply and H.E.P.

262a Name the three longest rivers that flow entirely within Wales.

262b What are the two main uses of water in Wales? Give two examples of each and name the rivers involved.

262c What advantages favour hydro-electric power production and water storage in Wales? (Think in terms of rainfall, geology, gradient, suitable sites, available land, etc.)

262d What other form of power station is now operating in North Wales? Name the example in Anglesey.

Fig. 263 Population density

263a Where are (i) the highest, and (ii) the lowest population densities in Wales?

263b Explain the low population density in Central Wales.

263c Are more, or less, than 50% of the people in the most populated areas able to speak Welsh?

261 Main agricultural areas

Dairying
Forestry
Stock raising
Mountain sheep farming
(after H. J. Savoury)

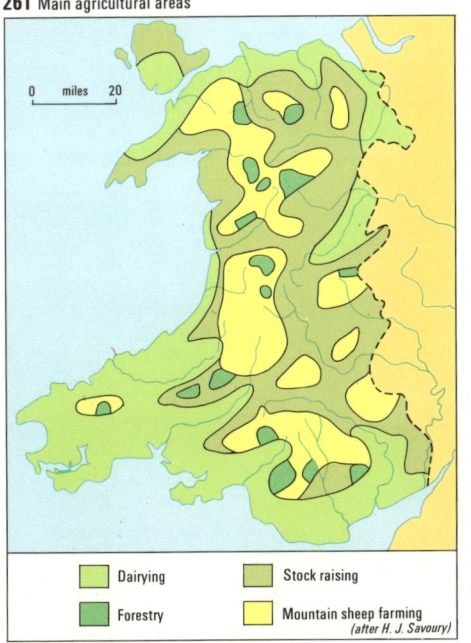

262 Water supply and Hydro-electric power

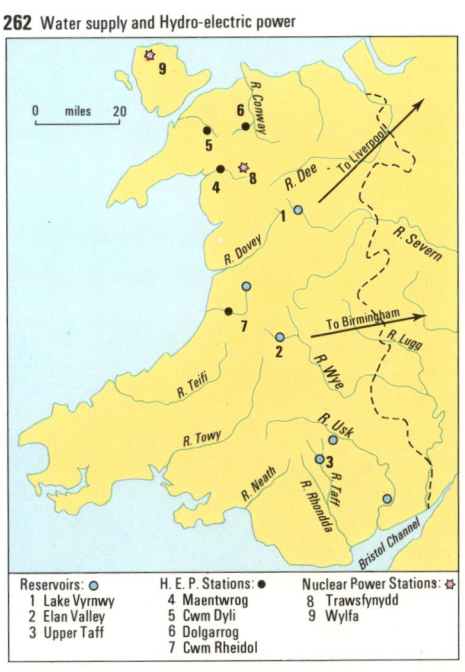

Reservoirs:
H.E.P. Stations:
Nuclear Power Stations:
1 Lake Vyrnwy
2 Elan Valley
3 Upper Taff
4 Maentwrog
5 Cwm Dyli
6 Dolgarrog
7 Cwm Rheidol
8 Trawsfynydd
9 Wylfa

263 Population density

More than 600 persons per sq. mile
50-600 persons per sq. mile
10-50 persons per sq. mile
Less than 10 persons per sq. mile
Area within which 50% of people are able to speak Welsh

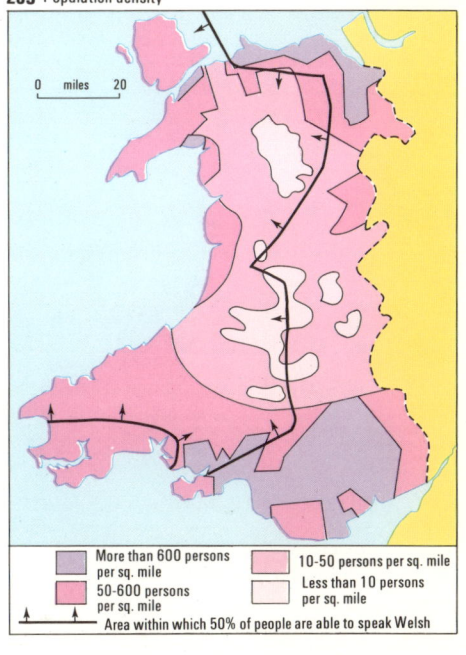

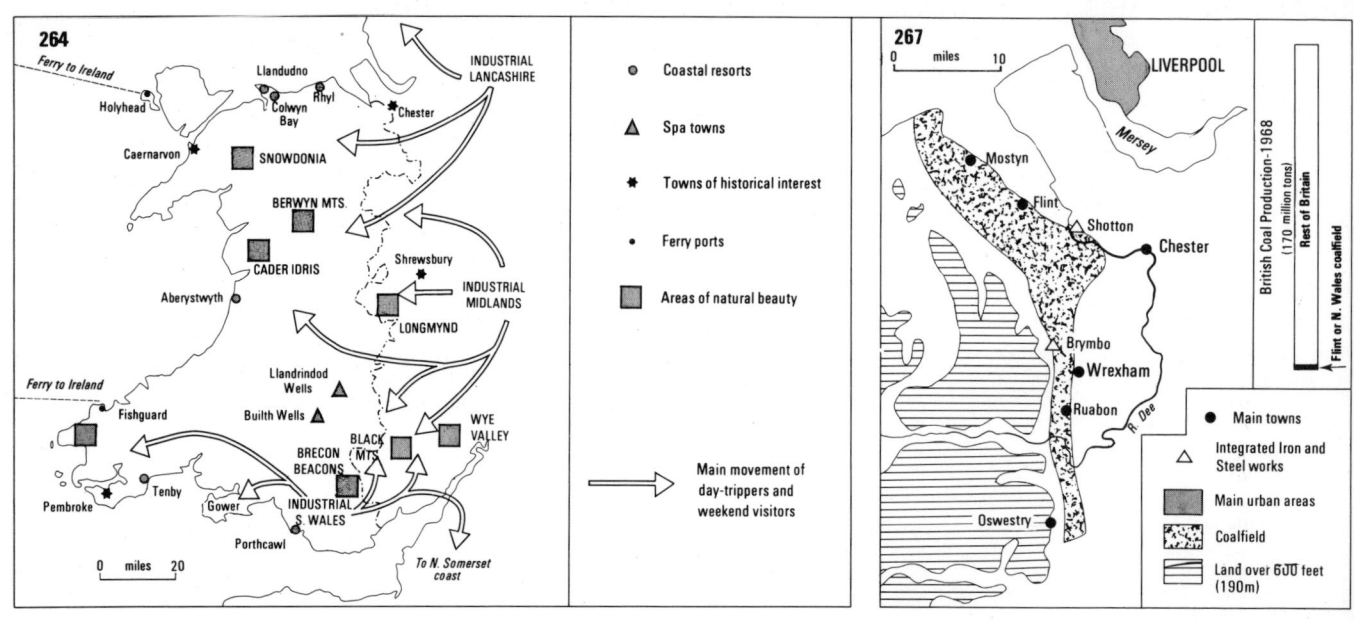

WALES Tourism, Mining and Quarrying

Fig. 264 Tourism

Wales has a considerable variety of coastal scenery, a glaciated upland interior, and a climate which is mild and wet in the winter and warm and relatively dry in the summer.

264a Name one resort on each of the northern, western and southern coasts of Wales.

264b (i) What other attractions has Wales for the tourist?
(ii) What is a 'spa town'?

264c Where do most of the tourists come from—particularly day-trippers and week-end visitors?

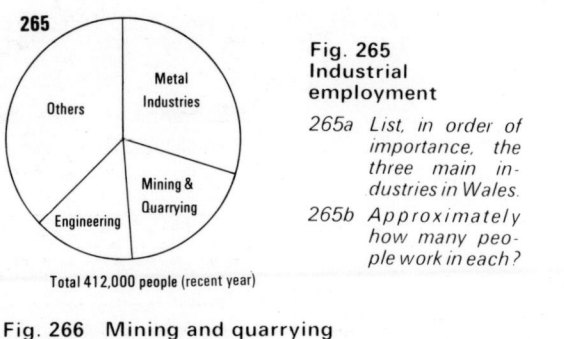

Fig. 265 Industrial employment

265a List, in order of importance, the three main industries in Wales.

265b Approximately how many people work in each?

Fig. 266 Mining and quarrying

The two coalfields (North Wales and South Wales) engage most of the workers in mining and quarrying. Minerals (indicated on Fig. 266) occur within and next to the volcanic and metamorphic rocks of North and Central Wales.

266a (i) Name three rocks, other than coal, mined or quarried in Wales.
(ii) Indicate their uses.

266b Locate and name the main Welsh slate quarries.

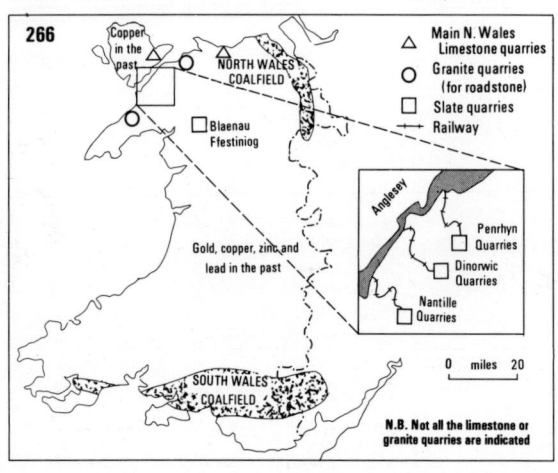

Fig. 267 North Wales coalfield

267a (i) Indicate the length in miles of the North Wales coalfield, and
(ii) describe its shape and position.

267b Is this coalfield an important contributor to total British coal production?

267c Name (i) one important industry on this coalfield other than coalmining, and (ii) its two main centres.

267d Name two other towns on or near the coalfield.

The variety of industry on this coalfield is considerable (e.g. artificial fibres, chemicals, engineering, paper, etc.), but this is one of the least important British coalfield areas.

Fig. 268 South Wales: comparative population density

268a In what part of Wales do most people live?

268b Approximately how many people live in Industrial South Wales?

268c Is the density of population high or low in South Wales? (Check your answer with Fig. 263.)

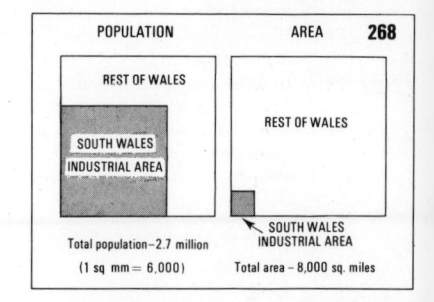

Fig. 269 South Wales: coal industry

269a How many million tons of coal are produced in the South Wales coalfield each year?

269b Are most of the main collieries in the east or west of the coalfield?

269b Most of the collieries in the eastern part of the coalfield are in valleys or on valley sides. Name the two main rivers in this part of the coalfield.

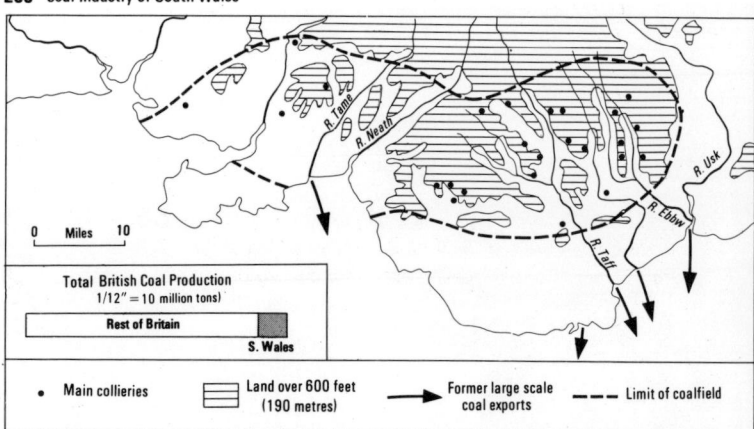

SOUTH WALES Industry

COALMINING, IRON AND STEEL, TINPLATING AND METAL REFINING

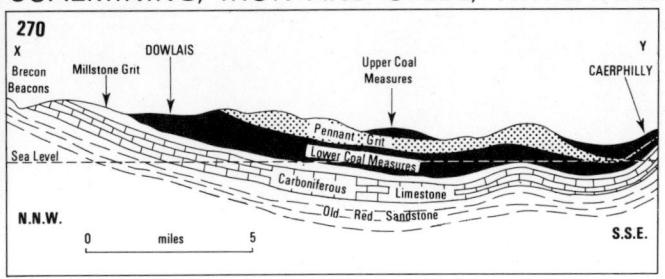

Fig. 270 Cross-section of the South Wales coalfield

The line of the cross-section is indicated on Fig. 271.
Coal used to be exported in large quantities to countries without other forms of power, and to world-wide steamship refuelling stations. Steam locomotives also devoured vast quantities.

270a Explain why coalmining started in the extreme northern and southern parts of the coalfield.
270b What rocks overlie and underlie the coal measures in the centre of the coalfield?
270c An anticline is strata (layers of rock) uplifted to form an arch, a syncline is strata downfolded to form a trough. Which word describes the South Wales coalfield?
270d Explain the decline of the coal industry in South Wales.

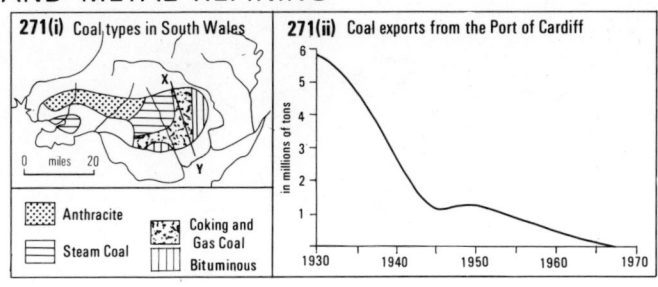

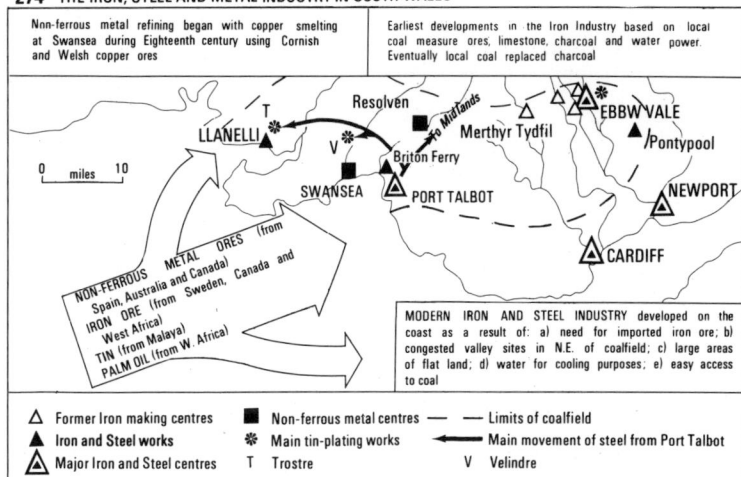

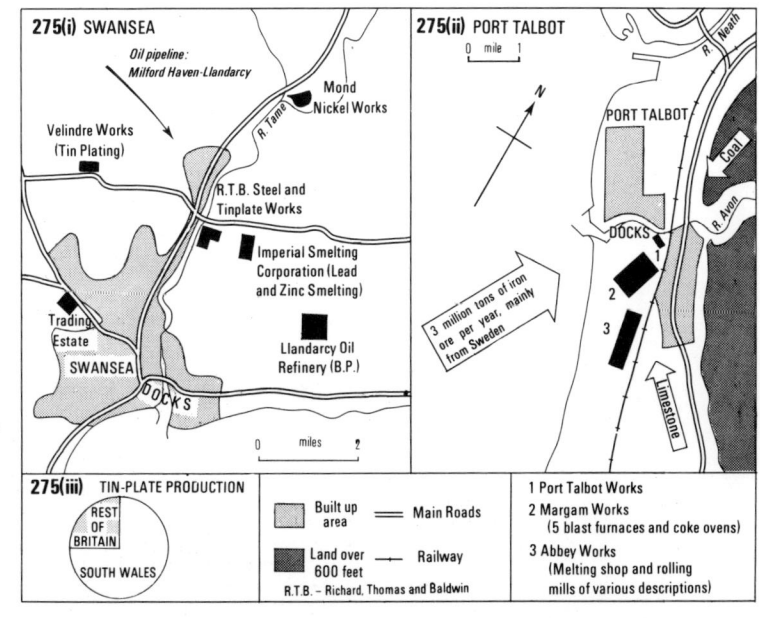

Fig. 271 Coal industry

271a Outline the main differences in the types of coal produced in the west and east of the coalfield.
271b What has happened to coal exports from the port of Cardiff since 1930?
271c See Fig. 274 and name three other ports that formerly exported large quantities of coal, and the types of coal concerned.

Fig. 272 Main uses of South Wales coal today

272a What are the two main uses of South Wales coal?
272b What industry consumes the coal converted in the coke ovens?

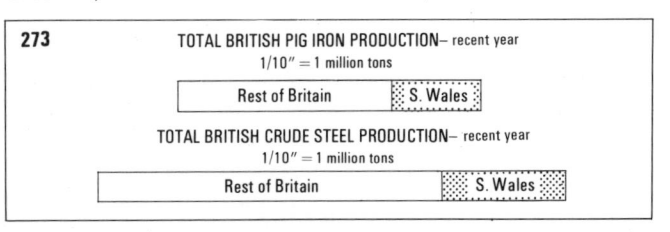

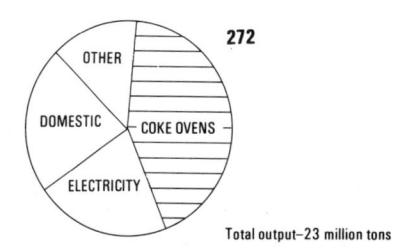

Fig. 273/274 Iron; steel industry

273/274a How many million tons of (i) pig iron, (ii) crude steel are produced in South Wales each year?
273/274b What percentage of total British (i) pig iron, (ii) crude steel is produced in South Wales?
273/274c In what part of the South Wales coalfield were the earliest iron works established, and why?
273/274d Where are most of the major iron and steel works today? Give two examples.
273/274e What advantages has a coastal site for this industry?

The only major iron and steel works not on the coast was maintained with government help because of the unemployment problems in the area.

273/274f Locate this iron and steel works.
273/274g A large proportion of the sheet steel produced at Port Talbot is sent to the tin-plating industry. Name the two main centres of this industry.
273/274h What percentage of total British tin-plate production comes from South Wales—see Fig. 275?
273/274i Where are the tin and palm oil, used in this industry, obtained?
273/274j What is the function of the palm oil?
273/274k Another important use of sheet steel is car bodies. What region receives Welsh steel for this purpose?
273/274l (i) Give two examples of a non-ferrous metal.
    (ii) In what town did the non-ferrous metal processing industry begin?
    (iii) What are the main countries supplying South Wales with non-ferrous metals?

Fig. 275 Swansea and Port Talbot

275a At the mouth of what river is Swansea situated?
275b What metals are refined or processed in the Swansea area?
275c What is the name of the oil refinery east of Swansea?
275d Port Talbot is one of the largest integrated iron and steel works in Europe.
    (i) Name its 3 main parts.
    (ii) Where does this works obtain its raw materials?

SOUTH WALES   New Industries; Main towns

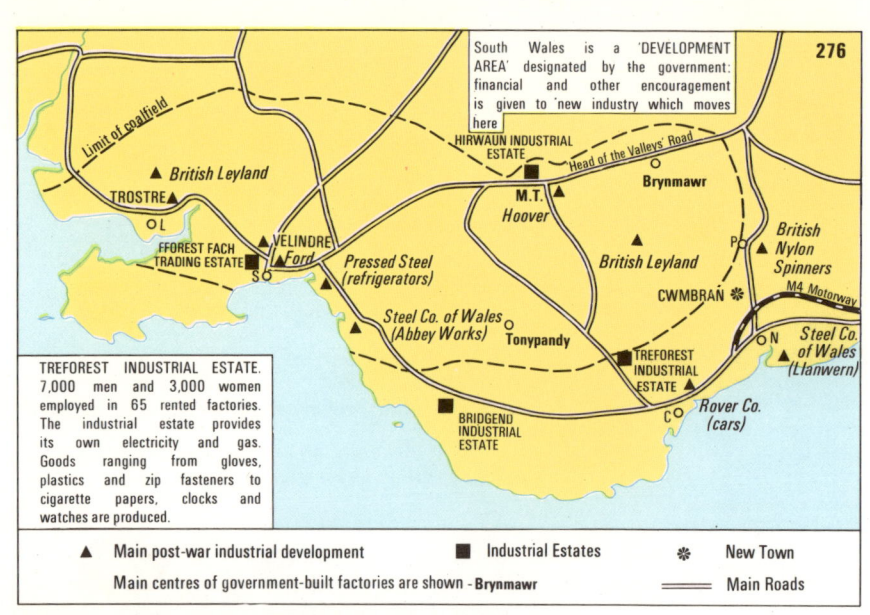

Fig. 276 Main industrial estates and new industries

Since 1930 South Wales has endured considerable unemployment through the decline of the coal industry, and it is now a Development Area.

276a (i) Explain 'Development Area'.
(ii) Name three towns where there are Government-built factories.
(iii) Name three industrial estates in South Wales.
(iv) Explain what you understand by an industrial estate.

276b (i) Give examples of a textile firm, a car firm, and a household appliance firm that have established works in South Wales since 1945.
(ii) Indicate where these works occur.

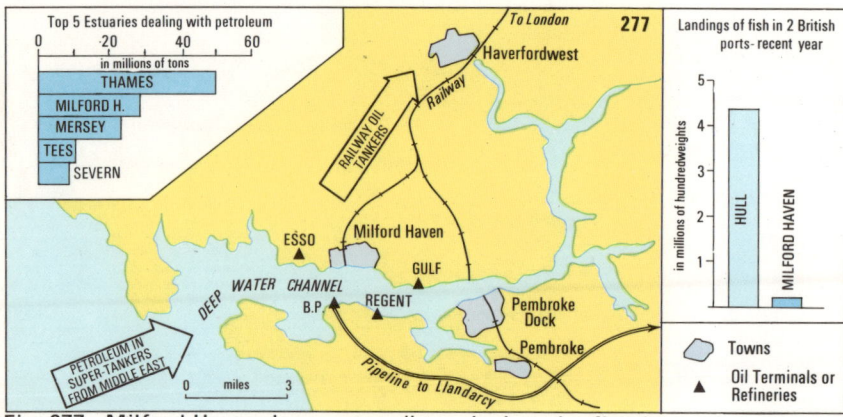

Fig. 277 Milford Haven deep water oil terminals and refineries

277a In what part of Wales is Milford Haven?
277b (i) How many million tons of petroleum are handled in Milford Haven each year?
(ii) Where is most of it obtained?
277c What advantages has Milford Haven for coping with super-tankers?

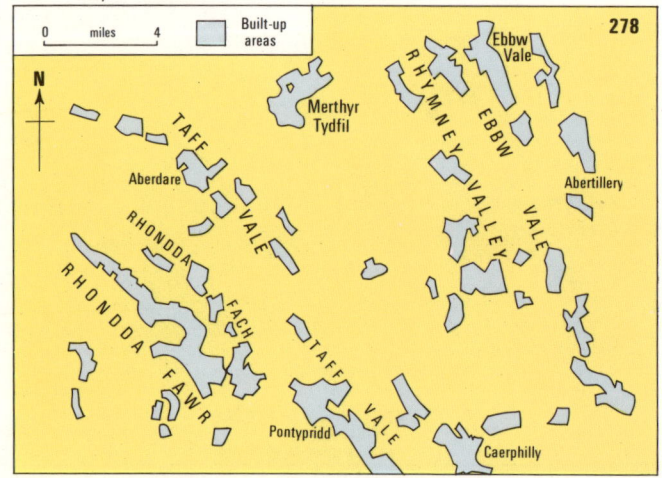

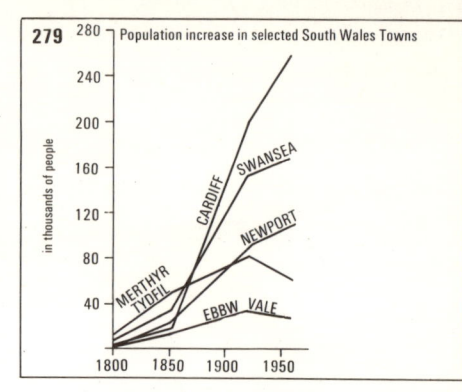

Fig. 279 Population growth

279a What was the largest town in South Wales in 1800?
279b List (i) the three largest towns today, (ii) the approximate population of each.
279c (i) During what period did Cardiff's most rapid growth take place?
(ii) How does this period of growth compare with that of Merthyr Tydfil?

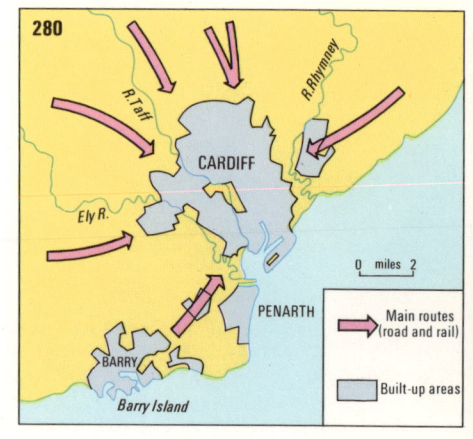

Fig. 280 Cardiff: site and situation

280a (i) At the mouth of what river is Cardiff situated?
(ii) What other rivers enter the Bristol Channel nearby?
280b Which of the following factors have contributed to the growth of Cardiff: route centre, seaport, coal and iron-ore mining centre, capital of Wales, geographical centre of the South Wales industrial area, important industrial town?

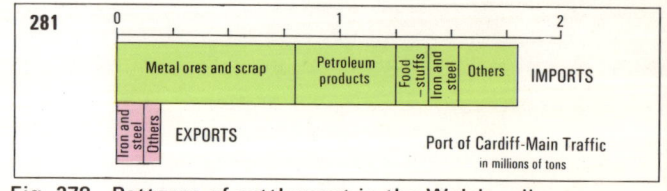

Fig. 278 Patterns of settlement in the Welsh valleys

278a What and where are the 'Welsh Valleys'?
278b What is the alignment (direction) of most of these valleys? (Give two examples.)
278c (i) What shape are most of the towns?
(ii) Suggest a reason for your answer.

Fig. 281 Cardiff: main port traffic

281a Is the import or export trade of Cardiff more important?
281b What are Cardiff's three main imports?
281c (i) Which of the following South Wales ports are likely to have the same sort of trade as Cardiff: Llanelly, Swansea, Port Talbot, Milford Haven, Newport?
(ii) Explain the exception.

SOUTH-WEST ENGLAND Physical 47

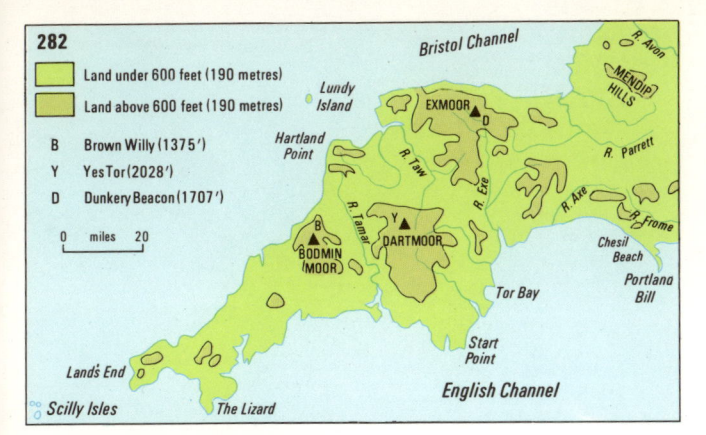

Fig. 282 Relief and drainage

282a Name, and indicate the main direction of flow, of the three longest rivers in South-West England.

282b (i) Name the highest peak in South-West England.
(ii) Indicate its height and position.

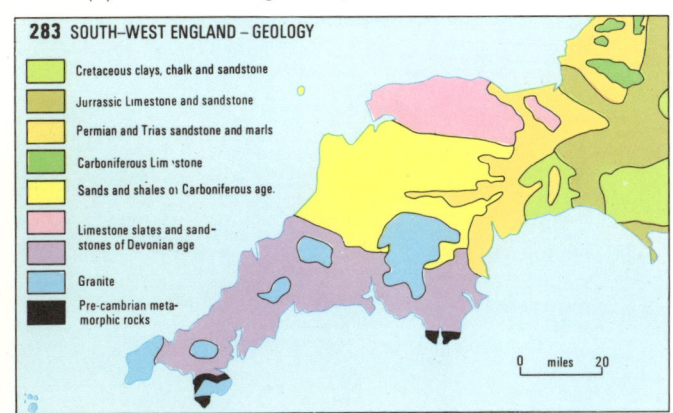

Fig. 283 Geology

283a What rock occurs both at the Lizard and at Start Point?

283b (i) Name the two main granite areas in South-West England.
(ii) Are they highland or lowland areas?

283c Name two other highland areas and indicate their geology.

Fig. 284 Rias and fjord coastlines

Rias are drowned river valleys, and are often confused with fjords ('lochs' in Scotland), which are drowned glaciated valleys. South-West England has a ria coastline.

284a Carrick Roads is a ria. Find four others.

284b (i) Find the Sound of Jura on an atlas map of Scotland.
(ii) Distinguish three differences between rias and fjords. (Think of shape, islands, cross-sections, etc.) Use simple diagrams to illustrate your answer where possible.

Fig. 285 Wind rose for Penzance

285a From what direction do the three most common winds experienced at Penzance blow?

285b What ocean do these prevailing winds cross?

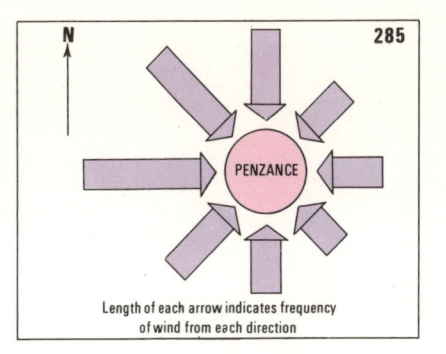

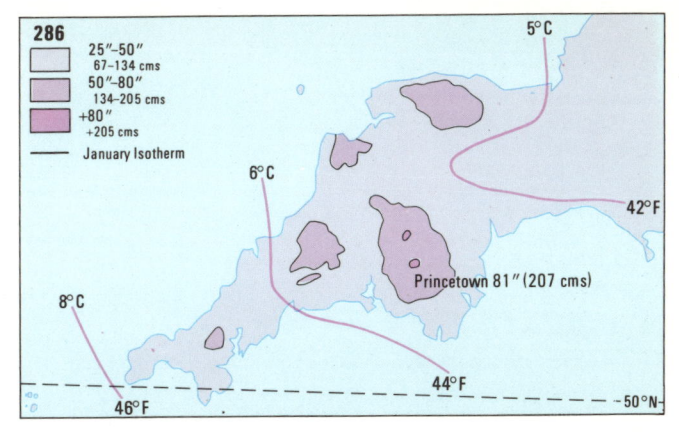

Fig. 286 Climate

286a Which parts of South-West England receive most rainfall?

286b (i) Name the place with the highest rainfall in South-West England.
(ii) How much rainfall does it receive each year?
(iii) How many inches of rainfall does most of South-West England receive?

286c (i) Which is the warmest part of South-West England in winter?
(ii) In what direction do winter temperatures decrease?

In winter, South-West England is one of the warmest regions in Britain; in summer it is one of the coolest.

286d Explain why its position facing the Atlantic is significant.

Fig. 287 Sunshine and frosts

287a Which parts of South-West England receive an average of more than 4½ hours of sunshine per day?

287b What is the significance of temperatures falling below 32°F (0°C.)?

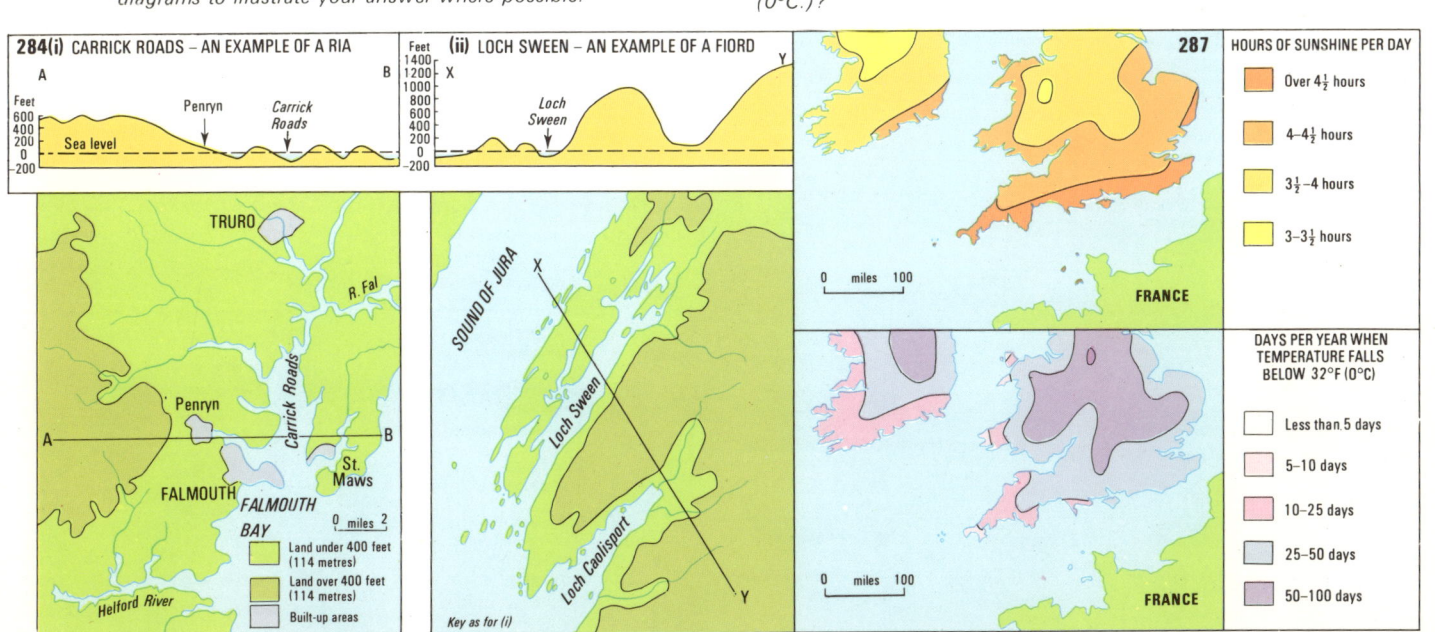

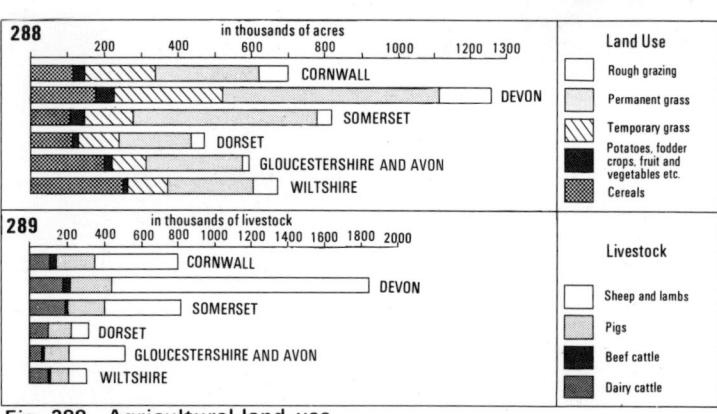

48 SOUTH-WEST ENGLAND Economy

Fig. 288 Agricultural land-use
(Gloucestershire-Avon-Wiltshire, are considered here for convenience.)

288a (i) What is the most important use of agricultural land in South-West England?
(ii) Calculate what percentage of Devon's agricultural land is pasture (i.e. temporary and permanent grass and rough grazing).

288b (i) Is pasture-land more, or less, important in Gloucestershire and Wiltshire?
(ii) Explain your answer (climate is important).

Fig. 289 Livestock
Dairying is the most important type of farming in South-West England.

289a (i) Name the two main dairying counties.
(ii) What advantages and disadvantages exist in South-West England for dairying? (Think in terms of rainfall, winter temperatures—i.e. is grazing possible all year round?— communications, markets, etc.)

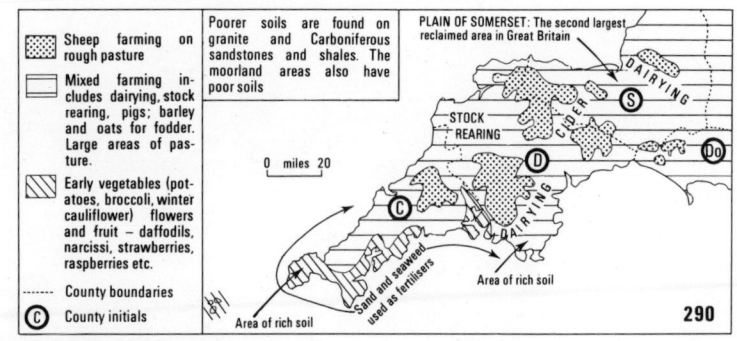

Fig. 290 Agriculture

290a Name three distinct areas where early vegetables, flowers and fruit are produced.

290b What factors favour the growing of early produce in South-West England? (Think in terms of winter temperatures, frosts, hours of sunshine, sheltered slopes, soil, etc.)

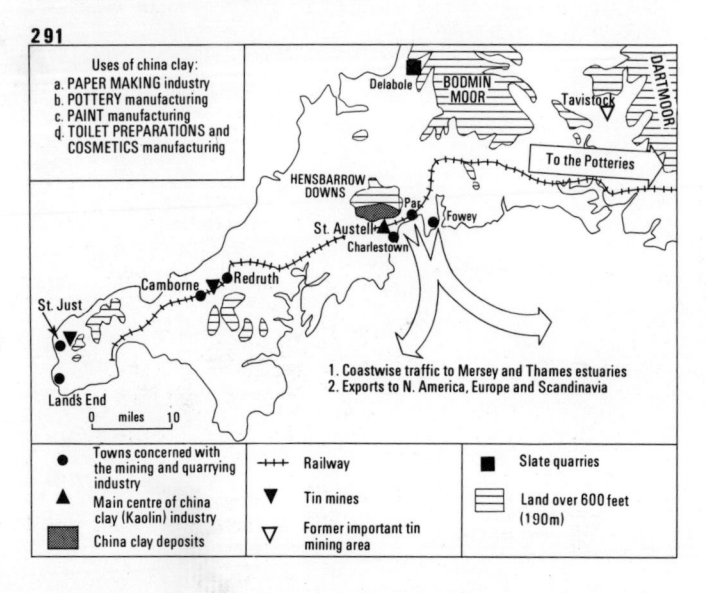

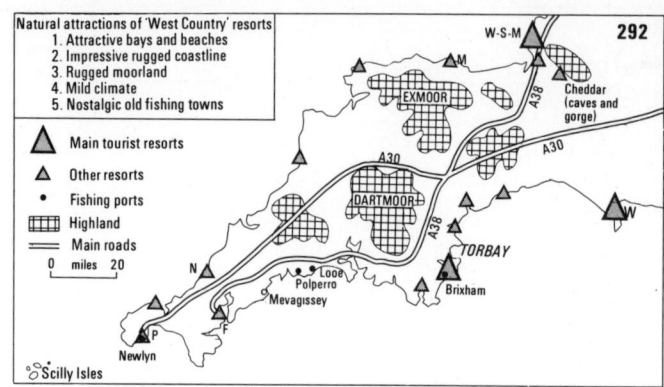

Fig. 292 Tourism

292a Name (i) the three main tourist resorts in South-West England, (ii) the three towns that make up Torbay, and (iii) three other resorts.

292b List some of the climatic and scenic attractions this area possesses.

Fig. 293 Main towns and industries

293a Name (i) the three main regional centres in South-West England, (ii) the market towns indicated by initials.

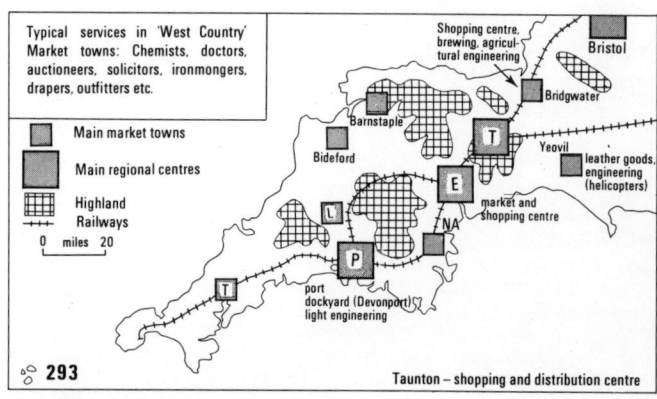

Fig. 291 Mining and quarrying

Copper and tin have been mined intermittently in Cornwall and West Devon since the Bronze Age, and in the nineteenth century this region was the leading world producer.

291a (i) How many tin mines still operate today?
(ii) Near what towns are they situated?

291b What rock contains the tin ore at the St Just mine? (See Fig. 283.)

The most valuable mineral mined in South-West England today is kaolin, and over half of it is exported.

291c (i) What is the alternative name for kaolin?
(ii) From what rock is kaolin derived (see Fig. 283)?

291d Indicate where kaolin is mined, and the main centre of this industry.

291e Name the three ports engaged in kaolin export.

291f (i) How is kaolin transported to the Potteries (Stoke-on-Trent)?
(ii) What is its use there? (See page 53.)

Fig. 294 Wells, Somerset: typical industries in a rural centre

294a Distinguish the industries in Wells that supply, or are supplied by, the local agricultural area.

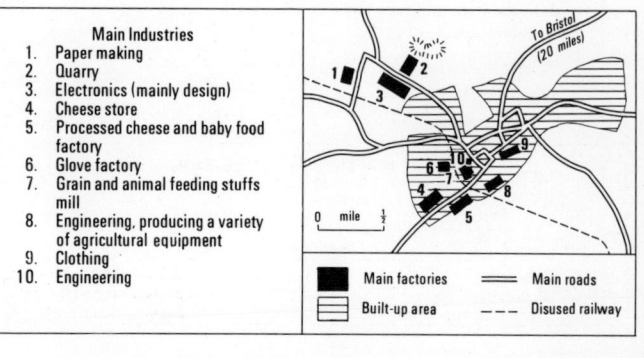

ENGLAND   The Bristol Region

295 The Bristol Region: Drainage, Relief and Industry

297 Bristol—Original site and City docks

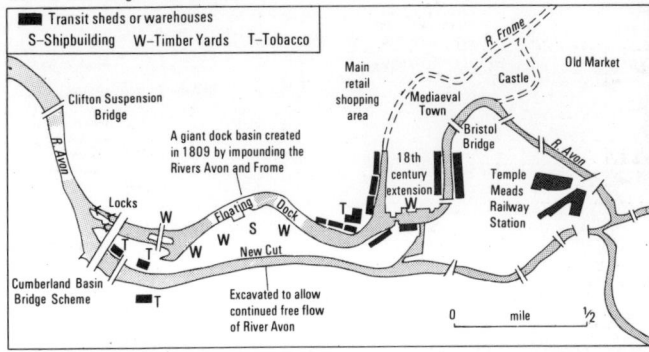

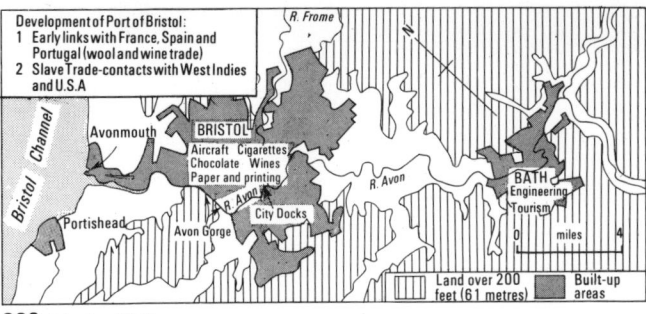

298 Bristol and Bath

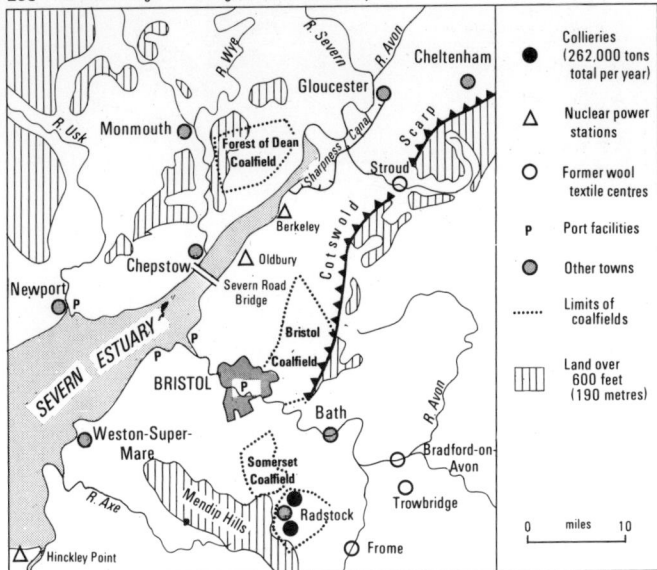

Fig. 295 Relief and Drainage

295a Name (*i*) four rivers that flow into the Severn estuary, (*ii*) the main towns through which they flow.

Dairying is important in South Avon and in the lowland areas between Bristol and Cheltenham. Name these, see Fig. 299.

295b Suggest (*i*) what towns are supplied with milk by these areas, (*ii*) what advantages the areas have for dairying.

Meat production (lamb and beef) and arable farming have replaced sheep farming (for wool) on the Cotswolds. A wool textile industry based on local supplies of wool and waterpower was formerly important.

295c (*i*) Name the old woollen centres.
(*ii*) **What advantage had West and South Yorkshire after the Industrial Revolution?** (See page 34)

295d (*i*) List the three small coalfields found in this region.
(*ii*) Which is still productive?
(*iii*) What other form of power is produced in this region, and where?

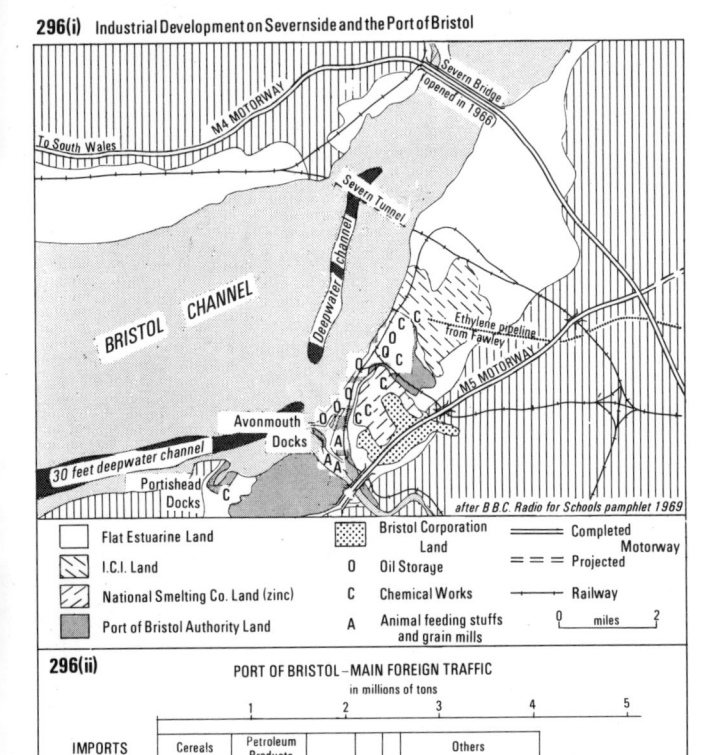

296(i) Industrial Development on Severnside and the Port of Bristol

296(ii) PORT OF BRISTOL—MAIN FOREIGN TRAFFIC

Figs. 296, 297 and 298 Bristol

The Port of Bristol is in three units. The part within Bristol itself is called the City Docks.

296a (*i*) Name the other two parts.
(*ii*) Which is the largest?

296b (*i*) List the three main foreign imports into the Port of Bristol.
(*ii*) How many times greater are imports than exports?
(*iii*) Explain this unbalance (think of Bristol's hinterland).

296c (*i*) What are the three main types of industry on Severnside?
(*ii*) Why are they located there?

296d (*i*) When was the Severn Road Bridge opened?
(*ii*) Calculate the mileage saving on a road journey from Bristol to Newport (see Fig. 295).

297a Explain the term 'floating dock'.

297b What cargo is handled at the City Docks?

297c Describe the original site of Bristol.

298a (*i*) How far from the Bristol Channel was the original site?
(*ii*) Suggest why Bristol grew up here and not at the present site of Avonmouth (think of shelter and protection).

298b What industries do you associate with the following well-known Bristol firms: Fry's, W.D. & H.O. Wills, Harvey's, B.A.C.?

298c Which of these industries arose out of early trade links?

298d What are Bath's tourist attractions?

Bristol is more than a seaport and industrial centre: it is the regional shopping, commercial and cultural centre of the South-West—and increasingly of South Wales too.

Fig. 299 Gloucester and Cheltenham

299a Explain Gloucester's importance as a route centre.

299b What type of industry occurs in the Gloucester area? Quote examples.

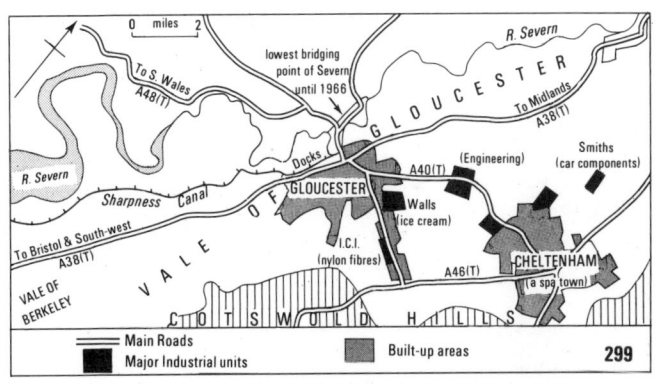

ENGLAND The Midlands

300 GEOLOGY

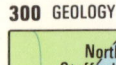

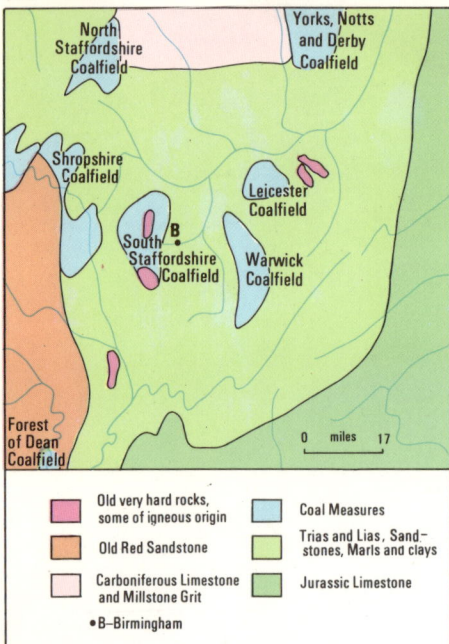

301 RELIEF AND DRAINAGE

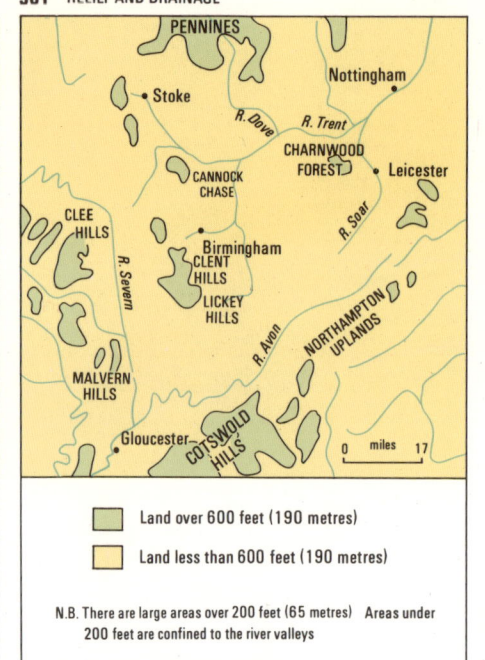

302 TEMPERATURE

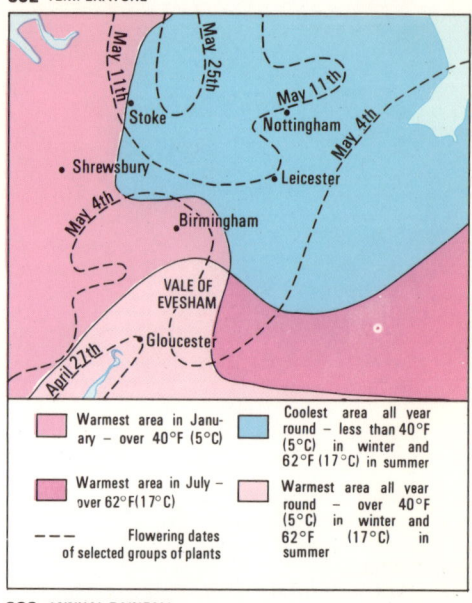

Key for 300:
- Old very hard rocks, some of igneous origin
- Old Red Sandstone
- Carboniferous Limestone and Millstone Grit
- Coal Measures
- Trias and Lias, Sandstones, Marls and clays
- Jurassic Limestone
- B—Birmingham

Key for 301:
- Land over 600 feet (190 metres)
- Land less than 600 feet (190 metres)

N.B. There are large areas over 200 feet (65 metres) Areas under 200 feet are confined to the river valleys

Key for 302:
- Warmest area in January – over 40°F (5°C)
- Warmest area in July – over 62°F (17°C)
- Coolest area all year round – less than 40°F (5°C) in winter and 62°F (17°C) in summer
- Warmest area all year round – over 40°F (5°C) in winter and 62°F (17°C) in summer
- ---- Flowering dates of selected groups of plants

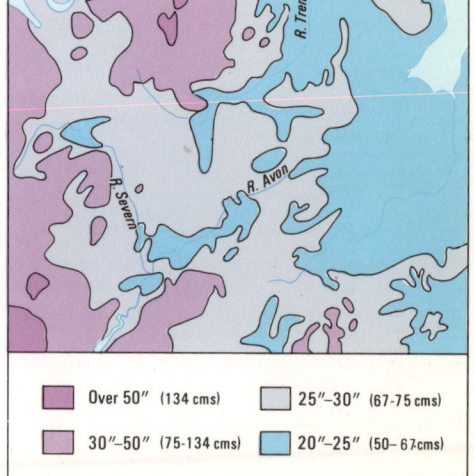

Figs. 300 and 301 Geology and relief

The Midlands is an area of Triassic and Liassic rocks.

300a (i) Which are the main types of rock?

(ii) What rocks bound the Midlands in the north, west, and south east?

300b Of what rock are (i) the Malverns, (ii) Charnwood Forest composed? (See Fig. 301.)

300c Map and name the Midland coalfields.

301a Name (i) the rivers, (ii) the hills that bound the Midlands in the north, west and south.

Fig. 302 and 303 Climate

302a Which part of the Midlands is (i) warmest all year round, (ii) coolest all year round?

302b (i) In what part of the Midlands are plants likely to flower and mature earliest?

(ii) Suggest why.

303a In which direction do annual rainfall totals decrease? Qualify your answer.

303b Explain why the valleys of the Avon and Trent receive low rainfall.

Fig. 304 Counties

There is now a new Metropolitan County known as West Midlands which is centred on Birmingham.

It includes part of S. Staffs and W. Warwickshire. Rutland is now part of Leicestershire. Herefordshire and Worcestershire form a joint county.

305b (i) Which two Midland counties have most land devoted to fruit and vegetables?

(ii) Name one of the areas concerned (see Fig. 302).

(iii) Explain why fruit and vegetable production has achieved so much importance in these areas.

Apples, plums, asparagus and brussels sprouts are the main products.

305c Why has Derbyshire most rough grazing land?

304 COUNTIES

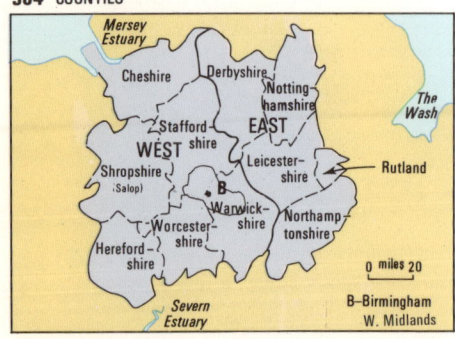

Figs. 305 and 306 Land use

305a Describe and account for the differences in agricultural land use in Herefordshire and Northamptonshire (see Figs. 301–304)

303 ANNUAL RAINFALL

Key:
- Over 50" (134 cms)
- 30"–50" (75-134 cms)
- 25"–30" (67-75 cms)
- 20"–25" (50-67 cms)

306a (i) What are the main Midland dairying counties?

(ii) Are they East or West Midland counties?

(iii) What factors favour dairying in these counties? (Think of climate, soils and markets.)

306b (i) Which three counties are important for sheep farming?

(ii) Explain why. (See Figs. 300, 301 and 303.)

306c Is there any link between large numbers of livestock and a large proportion of grassland in a county? Give two examples.

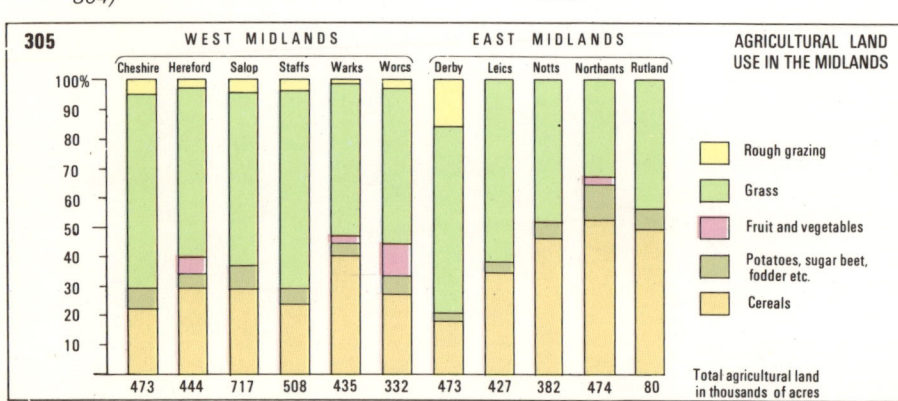

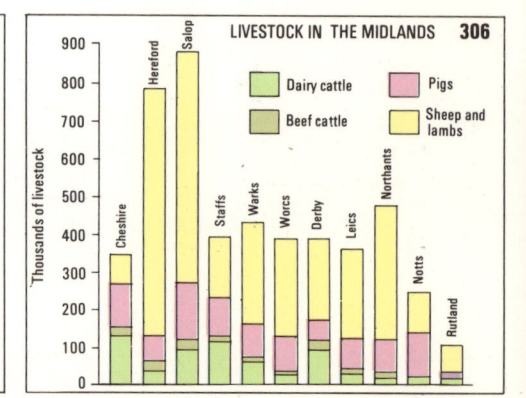

305 AGRICULTURAL LAND USE IN THE MIDLANDS

WEST MIDLANDS: Cheshire, Hereford, Salop, Staffs, Warks, Worcs
EAST MIDLANDS: Derby, Leics, Notts, Northants, Rutland

Categories:
- Rough grazing
- Grass
- Fruit and vegetables
- Potatoes, sugar beet, fodder etc.
- Cereals

Total agricultural land in thousands of acres: 473, 444, 717, 508, 435, 332, 473, 427, 382, 474, 80

306 LIVESTOCK IN THE MIDLANDS

Counties: Cheshire, Hereford, Salop, Staffs, Warks, Worcs, Derby, Leics, Northants, Notts, Rutland

Key:
- Dairy cattle
- Beef cattle
- Pigs
- Sheep and lambs

(Thousands of livestock)

ENGLAND The Midlands

307 The Shropshire Coalfield (N.B. Shropshire is now Salop)

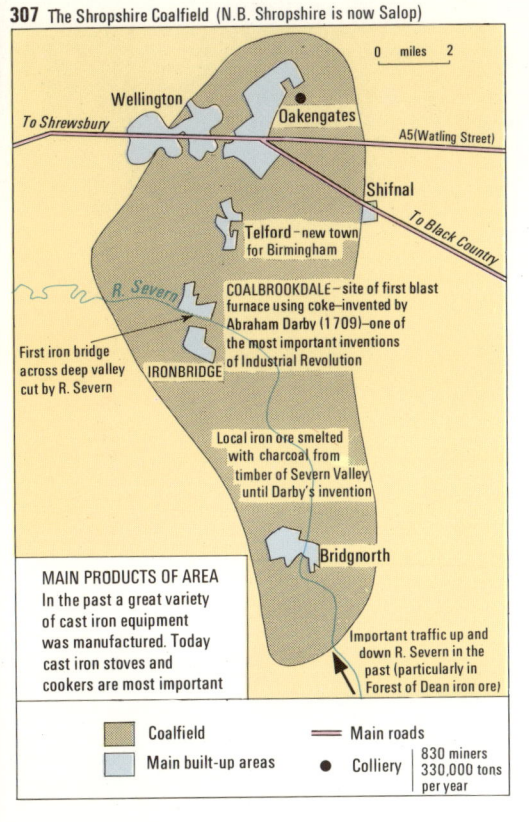

(Map showing: Wellington, Oakengates, To Shrewsbury, A5 (Watling Street), Shifnal, To Black Country, Telford – new town for Birmingham, R. Severn, COALBROOKDALE – site of first blast furnace using coke – invented by Abraham Darby (1709) – one of the most important inventions of Industrial Revolution, IRONBRIDGE – First iron bridge across deep valley cut by R. Severn, Local iron ore smelted with charcoal from timber of Severn Valley until Darby's invention, Bridgnorth, Important traffic up and down R. Severn in the past (particularly in Forest of Dean iron ore))

MAIN PRODUCTS OF AREA
In the past a great variety of cast iron equipment was manufactured. Today cast iron stoves and cookers are most important

Legend: Coalfield; Main built-up areas; Main roads; Colliery — 830 miners, 330,000 tons per year

309 Coal Production – recent year

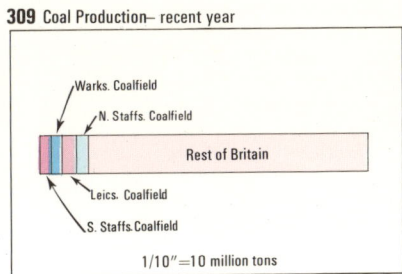

Warks. Coalfield / N. Staffs. Coalfield / Rest of Britain / Leics. Coalfield / S. Staffs. Coalfield

1/10" = 10 million tons

309a What proportion of total British coal production comes from the Midland coalfields?

310 Employment in main industries in the West Midlands (including Coventry and Stoke) – recent year

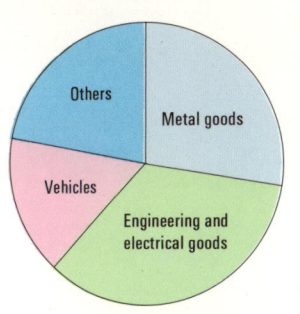

(Pie chart: Others, Metal goods, Vehicles, Engineering and electrical goods)

Total numbers employed – 1,215,000

310a List the three main West Midlands industries, in order of importance.

311(i) Employment in the Vehicle Industry – Main Regions – recent year

S. E. ENGLAND | WEST MIDLANDS (including Coventry and Stoke) | N. W. ENGLAND | OTHERS

1 centimetre = 100,000 employees

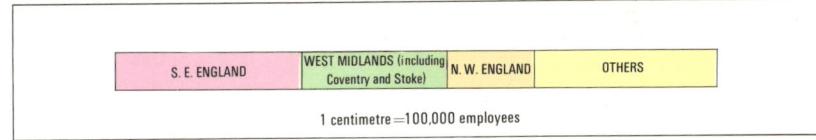

311a How does employment in the West Midlands (i) vehicle industry, (ii) metal goods industry, compare with other regions of Britain?

311(ii) Employment in the metal Goods Industry – Main Regions – recent year

Y. & H. | S. E. | WEST MIDLANDS (incl. Coventry and Stoke) | OTHERS

1 centimetre = 200,000 employees

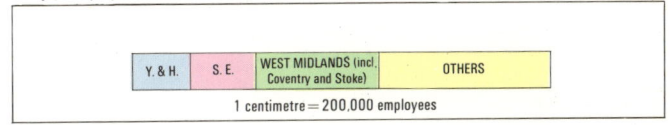

Figs. 307–312 Coalmining, metal-working

Today the Shropshire coalfield is the least important of the Midland coalfield areas.

307a (i) What is its historical significance?
(ii) What industries have survived today?

308a How does the South Staffordshire coalfield compare with the Shropshire coalfield in (i) size, and (ii) production?

308b (i) In what part of the S. Staffs. coalfield and (ii) near what town are the only remaining collieries?

308c Suggest two factors that favoured the development of an iron industry in South Staffordshire.

308d Where is the only integrated iron and steel plant today?

312a What is the alternative name for the South Staffordshire coalfield area?

312b List the main Black Country towns in order of size (see Fig. 308).

312c Is Birmingham part of the Black Country?

312d Show how the word conurbation applies to the West Midlands.

Iron and steel making has been replaced in the Black Country by iron and steel using. Most metal goods industries remain small, and there is still a tendency for one town to specialise in one type of product.

312e Give three examples of town specialisation.

312f Where are the nearest sources of (i) iron and steel, (ii) non-ferrous metals, for this industry? (See pages 45–46.)

312g What non-metal industries occur at (i) Walsall, (ii) Aldridge, (iii) Smethwick?

308 South Staffordshire Coalfield

(Map showing: POPULATION circles 150,000 / 120,000 / 97,000 / 64,000; Rugeley Power Station (supplied by conveyor belt from Lea Hall); Lea Hall (a new mine producing 1½ million tons per year); CANNOCK; IRON ORE formerly mined from coal measures; WALSALL; WOLVERHAMPTON; Bilston; Wednesbury; DUDLEY; WEST BROMWICH; Brierley Hill; BIRMINGHAM; Rapid decline of coalmining in southern part of coalfield since 1920)

Legend: Exposed coal measures; Limestone; Collieries; Collieries closed since 1965; Integrated iron and steel works; Steel works and rolling mills

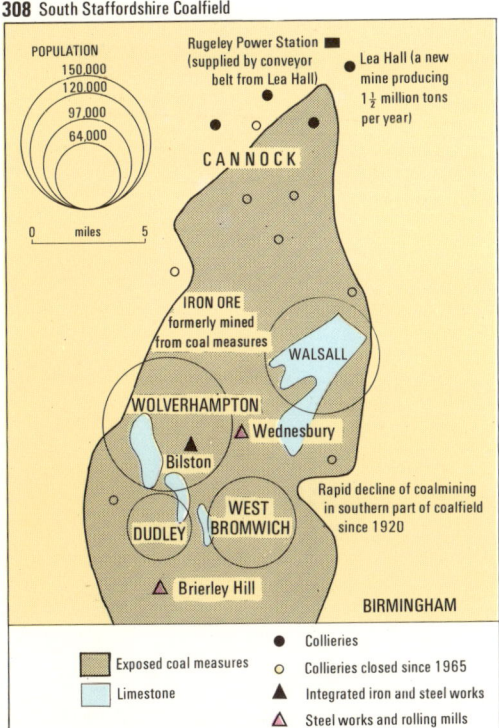

312 The West Midlands Conurbation

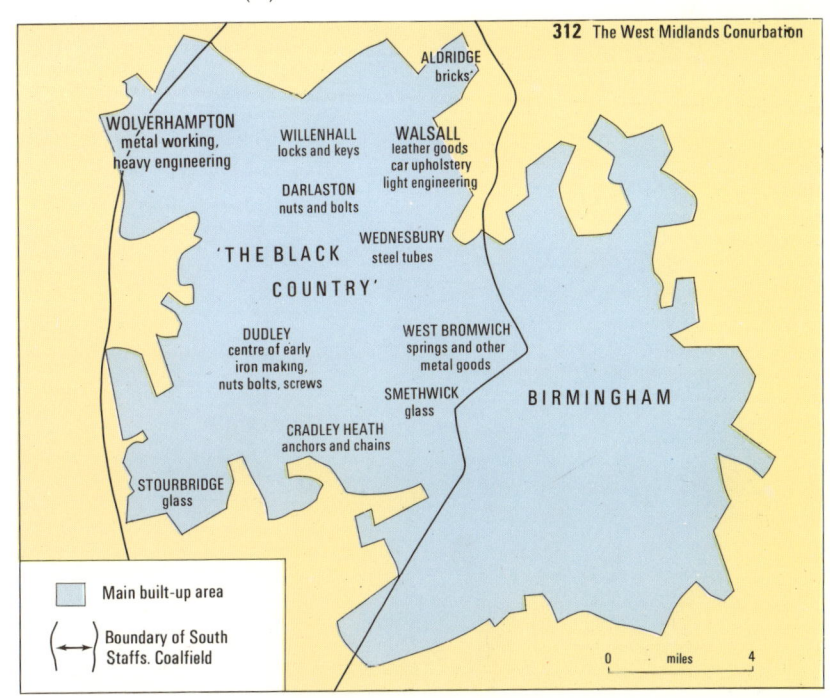

(Map showing: ALDRIDGE bricks; WOLVERHAMPTON metal working, heavy engineering; WILLENHALL locks and keys; WALSALL leather goods, car upholstery, light engineering; DARLASTON nuts and bolts; 'THE BLACK COUNTRY'; WEDNESBURY steel tubes; DUDLEY centre of early iron making, nuts bolts, screws; WEST BROMWICH springs and other metal goods; SMETHWICK glass; CRADLEY HEATH anchors and chains; STOURBRIDGE glass; BIRMINGHAM)

Legend: Main built-up area; Boundary of South Staffs. Coalfield

ENGLAND The Midlands

313 The growth of Birmingham

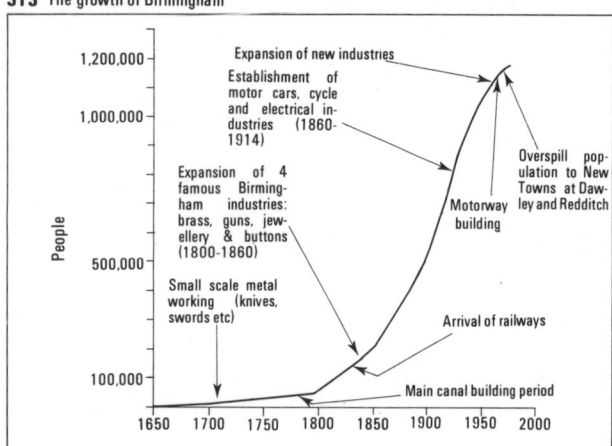

314 BIRMINGHAM – national route centre
(i) Canals
(ii) Railways and Motorways

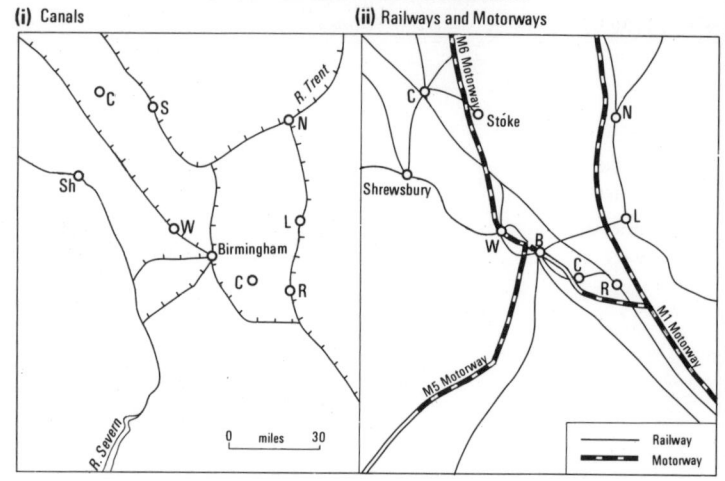

315 Birmingham: Industrial centre

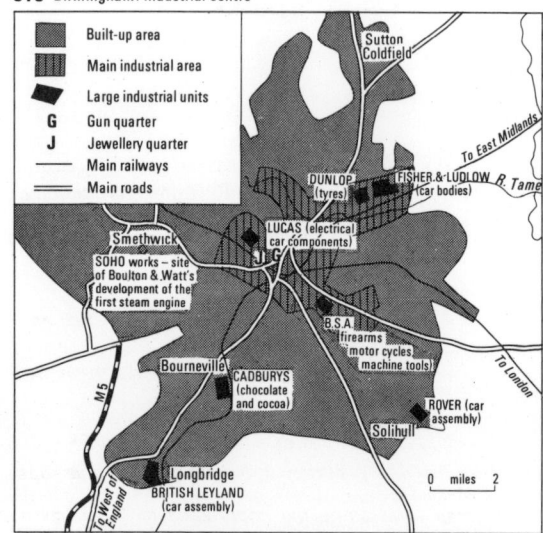

Figs. 313, 314 and 315 Birmingham

313a Describe and account for the growth of Birmingham since 1700.

314a Name the four estuaries and ports with which Birmingham has canal links.

314b (i) How important is canal transport today?
(ii) What types of cargo can still be transported cheaply, and effectively, by canal?

214c Describe Birmingham's position with regard to (i) the national railway network, (ii) the national motorway network.

314d (i) To what extent do Midland canals, railways and motorways follow similar routes?
(ii) What is the main direction of movement?

Birmingham is neither a seaport nor an inland river port, yet it has developed into, and has remained, a prosperous industrial city.

315a What industries were formerly important in Birmingham? (See also Fig. 313.)

315b Name (i) the two main car assembly firms in Birmingham (ii) the locations of their main works.

315c (i) What Birmingham firms are closely linked to this assembly industry?
(ii) Name the components supplied.

315d Name two other well-known Birmingham firms.

317a Which of the East Midlands coalfields is most productive today? (See Fig. 309.)

317b In what ways are Coventry's industries similar to those of Birmingham?

317c (i) What are the other main industries in the East Midlands?
(ii) Where do they occur?

317 Industry in the East Midlands

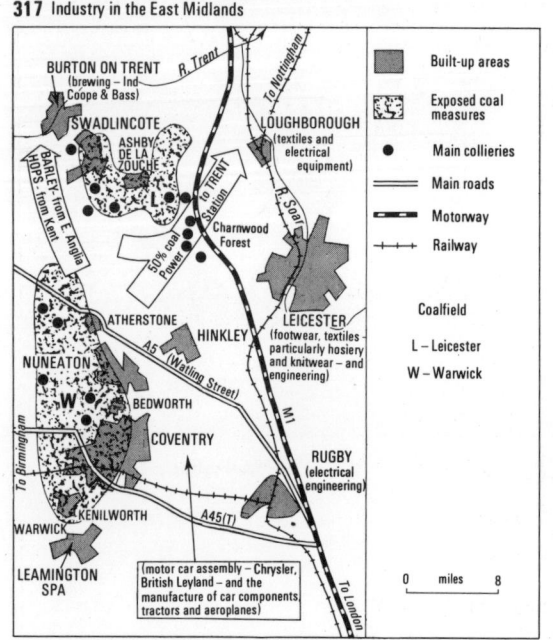

Figs. 316, 317 and 318 East Midlands mining, quarrying and the manufacturing industry

316a What are the two main manufacturing industries in the East Midlands?

318a (i) What proportion of British iron ore production comes from South Lincolnshire and Northamptonshire?
(ii) How much of this ore is retained locally?

318b (i) Name the main iron and steel centre in Northamptonshire.
(ii) What other towns have blast furnaces?

318c Name the four East Midland footwear towns.

318d The hides for this industry were once obtained locally. Where are they obtained today?

316 Employment in Manufacturing Industries in the East Midlands (including Derby and Nottingham)
Total numbers employed – 624,000

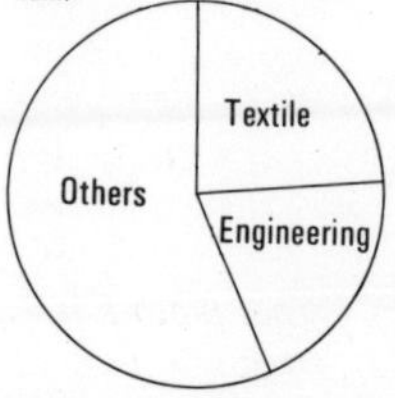

318 Iron and Steel in the East Midlands – recent year

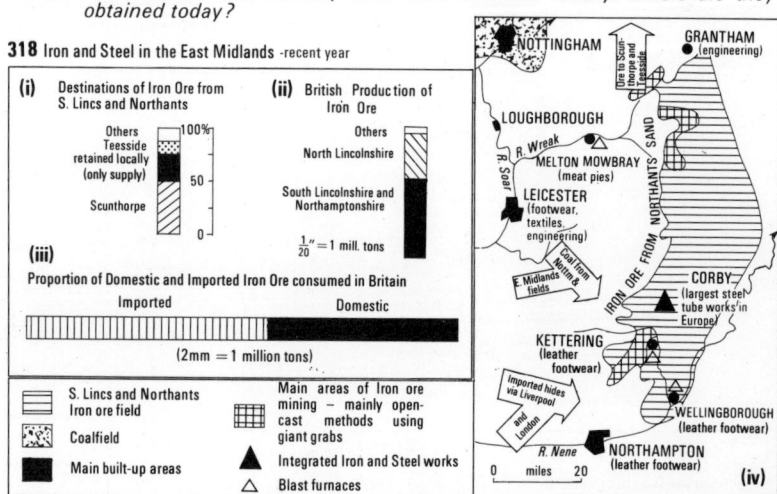

ENGLAND The Potteries

319 The North Staffordshire Coalfield (or Potteries)

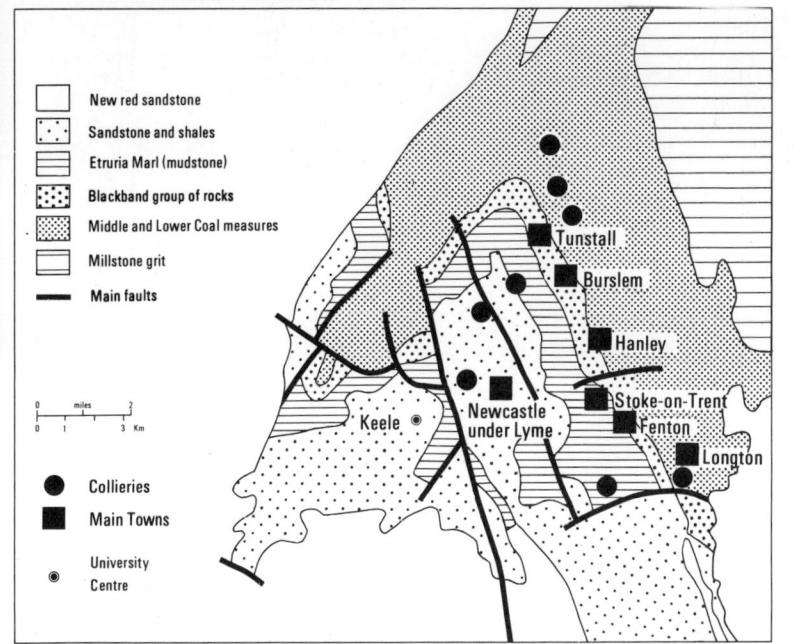

Legend:
- New red sandstone
- Sandstone and shales
- Etruria Marl (mudstone)
- Blackband group of rocks
- Middle and Lower Coal measures
- Millstone grit
- Main faults
- ● Collieries
- ■ Main Towns
- ◉ University Centre

Towns shown: Tunstall, Burslem, Hanley, Stoke-on-Trent, Fenton, Longton, Newcastle under Lyme, Keele

322 Raw materials for the Potteries

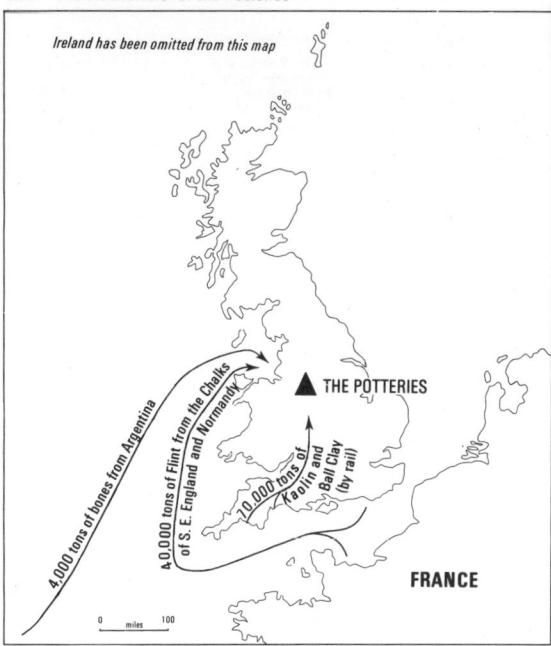

Ireland has been omitted from this map

- 4,000 tons of bones from Argentina
- 40,000 tons of Flint from the Chalks of S.E. England and Normandy
- 70,000 tons of Kaolin and Ball Clay (by rail)

THE POTTERIES — FRANCE

Figs. 319–323 Coalmining; pottery (ceramics) industry

319a (i) On the south-west corner of which hills are the Potteries situated?
(ii) Name three other coalfields that flank these hills.

319b (i) How many collieries are active today?
(ii) What is the annual output of this coalfield? (See Fig. 309.)

Keele and Newcastle-under-Lyme are separate towns. The remaining towns now form the new county borough of Stoke-on-Trent.

319c (i) Which are they? (ii) On what rock are most of them situated?

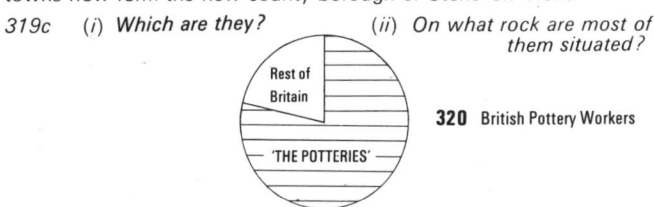

320 British Pottery Workers (Rest of Britain / 'THE POTTERIES')

Stoke is one of the main world centres of ceramics with an international reputation.

320a What proportion of British pottery workers is employed in North Staffordshire?

321a What factors led the local farmers to turn to pottery making?
321b What raw materials existed locally?

321 Development of the Pottery Industry

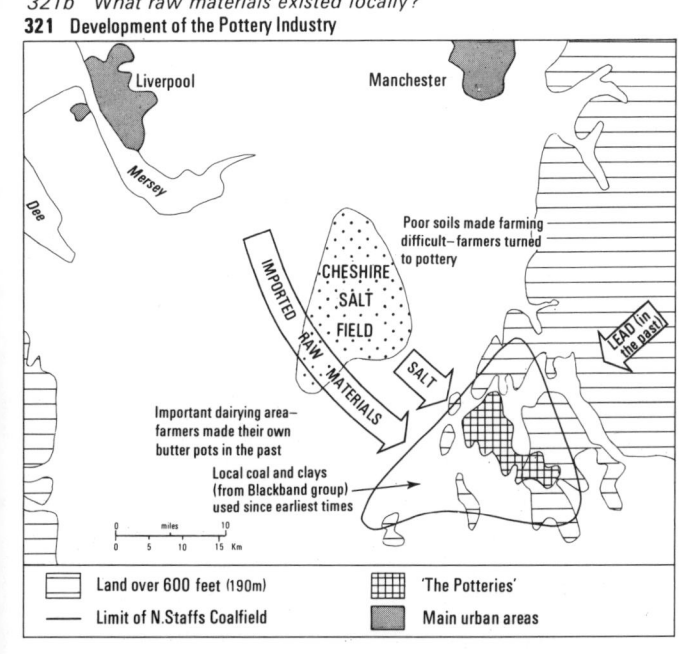

(Map showing Liverpool, Manchester, Mersey, Dee, Cheshire Salt Field, IMPORTED RAW MATERIALS, SALT, LEAD (in the past))

- Poor soils made farming difficult—farmers turned to pottery
- Important dairying area—farmers made their own butter pots in the past
- Local coal and clays (from Blackband group) used since earliest times

Legend: Land over 600 feet (190m); Limit of N.Staffs Coalfield; 'The Potteries'; Main urban areas

Pottery-making became a factory industry in the eighteenth century. From that time it has been in the hands of family concerns, e.g. Wedgwood's (1759), Minton (1793), and Copeland's (formerly Spode's).

322a (i) Where are the raw materials obtained today?
(ii) What forms of transport are used?

Today coal-fired bottle-ovens are being replaced by gas-fired and electric kilns.

322b (i) Do these changes mean a wholesale reduction in the demand for coal?
(ii) Explain your answer.

322c Why has a ceramics industry not developed in Cornwall?

323a (i) What was the first important link between the Potteries and the Mersey?
(ii) Draw a map to illustrate the importance of Crewe as a railway route centre.

The Potteries, besides producing a great variety of clay products (e.g. tableware, bricks, tiles, etc.) is also an iron and steel and engineering area.

323 Transport in N.W. England

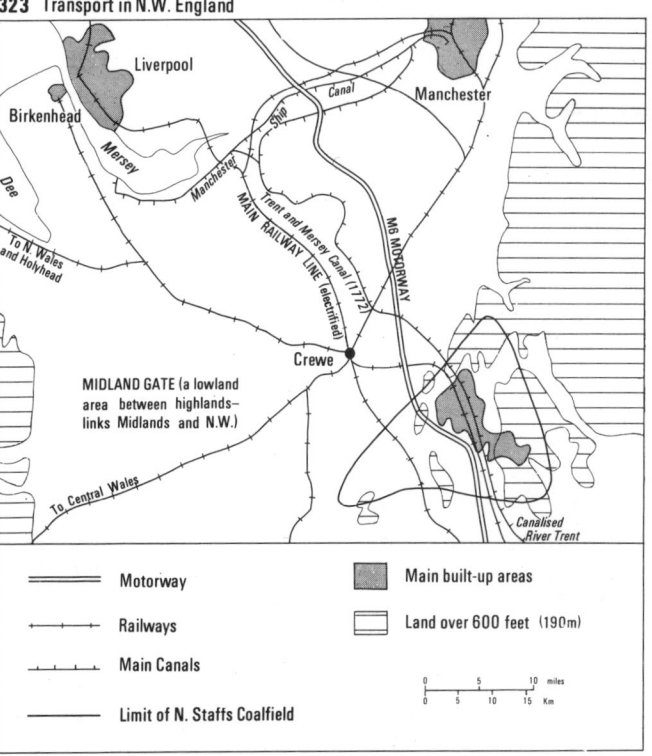

(Map showing Liverpool, Birkenhead, Manchester, Manchester Ship Canal, Mersey, Dee, Trent and Mersey Canal, MAIN RAILWAY LINE (electrified), M6 MOTORWAY, Crewe, Canalised River Trent)

MIDLAND GATE (a lowland area between highlands—links Midlands and N.W.)

To N Wales and Holyhead; To Central Wales

Legend: Motorway; Railways; Main Canals; Main built-up areas; Land over 600 feet (190m); Limit of N. Staffs Coalfield

EASTERN ENGLAND Physical and Agriculture

324 Geology and Soils in Eastern England

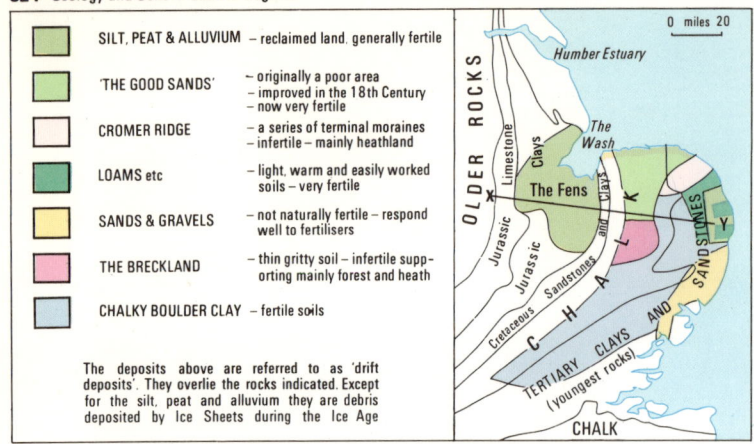

- SILT, PEAT & ALLUVIUM — reclaimed land, generally fertile
- 'THE GOOD SANDS' — originally a poor area – improved in the 18th Century – now very fertile
- CROMER RIDGE — a series of terminal moraines – infertile – mainly heathland
- LOAMS etc — light, warm and easily worked soils – very fertile
- SANDS & GRAVELS — not naturally fertile – respond well to fertilisers
- THE BRECKLAND — thin gritty soil – infertile supporting mainly forest and heath
- CHALKY BOULDER CLAY — fertile soils

The deposits above are referred to as 'drift deposits'. They overlie the rocks indicated. Except for the silt, peat and alluvium they are debris deposited by Ice Sheets during the Ice Age.

Figs. 324 and 325 Geology and relief

Eastern England includes several well-defined regions, the most important of which are (i) East Anglia (Norfolk and Suffolk—see Fig. 329) and (ii) The Fens (see Fig. 324).

324a List the sequence of rocks (from oldest to youngest in Eastern England.

324b (i) What are 'drift deposits'?
(ii) What are the two most widespread drift deposits in Eastern England?
(iii) In what counties are they most common? (See Fig. 329.)

325a What rocks make up (i) the Northampton Heights, (ii) the East Anglian Heights?

325 West-East cross section across Eastern England

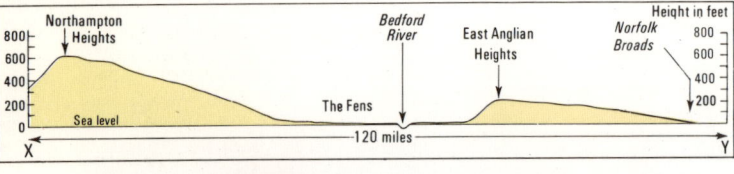

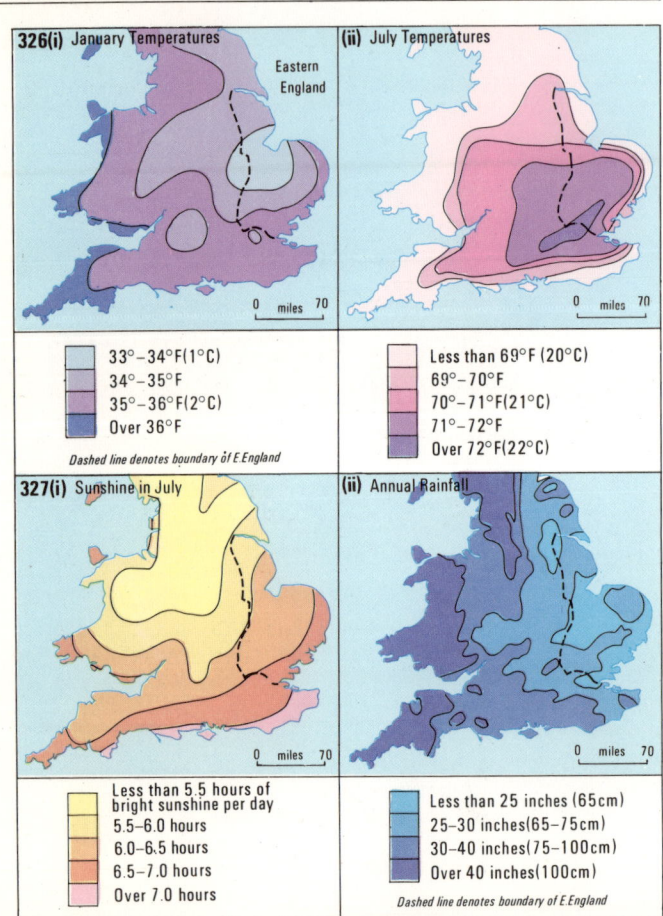

326(i) January Temperatures   (ii) July Temperatures

- 33°–34°F (1°C)
- 34°–35°F
- 35°–36°F (2°C)
- Over 36°F

Dashed line denotes boundary of E.England

- Less than 69°F (20°C)
- 69°–70°F
- 70°–71°F (21°C)
- 71°–72°F
- Over 72°F (22°C)

327(i) Sunshine in July   (ii) Annual Rainfall

- Less than 5.5 hours of bright sunshine per day
- 5.5–6.0 hours
- 6.0–6.5 hours
- 6.5–7.0 hours
- Over 7.0 hours

- Less than 25 inches (65cm)
- 25–30 inches (65–75cm)
- 30–40 inches (75–100cm)
- Over 40 inches (100cm)

Dashed line denotes boundary of E.England

328 Agricultural Land Use in Eastern England

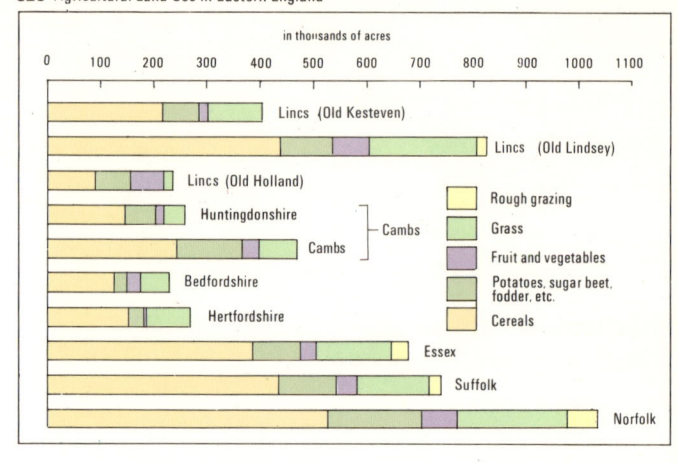

(in thousands of acres)

Lincs (Old Kesteven), Lincs (Old Lindsey), Lincs (Old Holland), Huntingdonshire, Cambs, Bedfordshire, Hertfordshire, Essex, Suffolk, Norfolk

- Rough grazing
- Grass
- Fruit and vegetables
- Potatoes, sugar beet, fodder, etc.
- Cereals

Figs. 328 and 329 Agriculture

328a (i) Is Eastern England mainly pasture or arable land?
(ii) Explain why—see Figs. 326 and 327.

328b (i) Which county has the largest acreage under cereals? (Give actual figure.)
(ii) Which county has the greatest proportion of total land under cereals?
(iii) List the three counties with the largest acreages under fruit and vegetables.

329a Which counties produce most (i) barley, (ii) wheat?

329b (i) What does the yield per acre indicate about the conditions under which a crop grows?
(ii) Which counties or parts of counties have the highest (1) barley, (2) wheat yields?

329c Which is the most important cereal crop in Eastern England?

329(i) Barley production in Eastern England   (ii) Wheat production in Eastern England

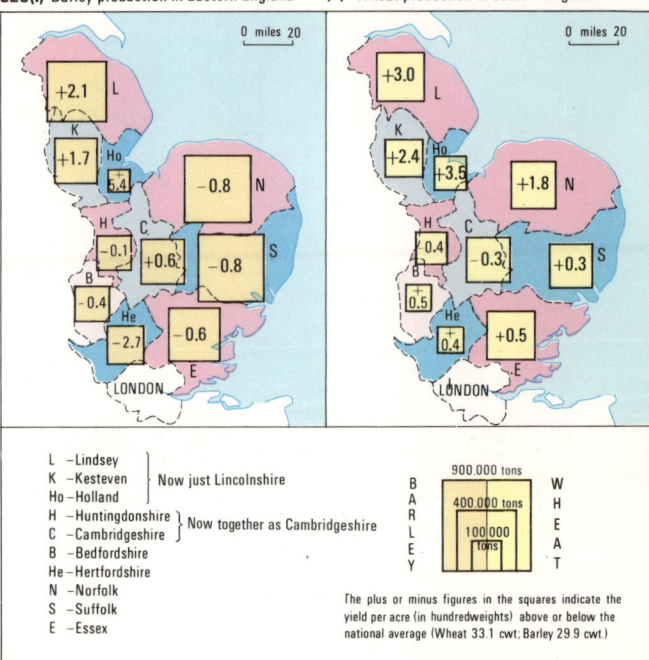

- L — Lindsey
- K — Kesteven } Now just Lincolnshire
- Ho — Holland
- H — Huntingdonshire } Now together as Cambridgeshire
- C — Cambridgeshire
- B — Bedfordshire
- He — Hertfordshire
- N — Norfolk
- S — Suffolk
- E — Essex

The plus or minus figures in the squares indicate the yield per acre (in hundredweights) above or below the national average (Wheat 33.1 cwt; Barley 29.9 cwt).

Figs. 326 and 327 Climate

326a (i) Which part of England and Wales is coldest in winter?
(ii) How cold?

326b (i) Which part of England and Wales is hottest in summer?
(ii) How hot?

326c Explain your answers to (a) and (b). (Think of differential rates of heating and cooling of land and sea, and their effects on air temperature and winds.)

327a (i) What is the average number of sunshine hours per day in Eastern England in July?
(ii) How does it compare with the rest of England and Wales?

327b Which parts of Eastern England have less than 25 inches of rainfall per year?

327c In what way are hours of sunshine and rainfall totals linked?

EASTERN ENGLAND Agriculture

330 Livestock in Eastern England

333 The Fens

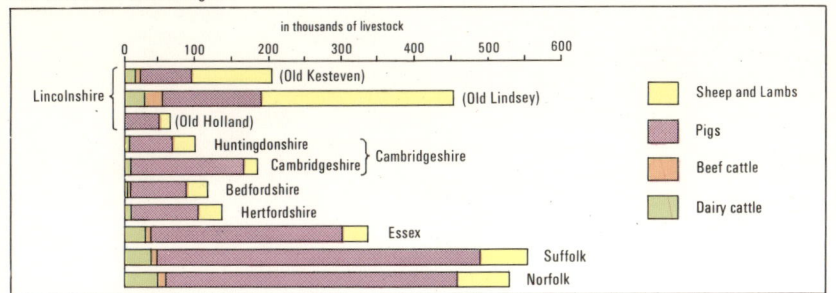

Figs. 330 and 331 Land use

330a (i) What are the most common livestock in Eastern England?
(ii) Are they fed on grain or grass?
(iii) In which three counties are they most important?

330b (i) Which county has most sheep?
(ii) Why? (See Fig. 324.)

330c Explain the high number of dairy cattle in Essex and Sussex. (Think of soils and market.)

331a What proportion of the national production of (i) ducks and (ii) poultry are reared in Eastern England; and (iii) which counties are most important?

331b Which counties have most land under (i) sugar beet, (ii) potatoes, (iii) bulbs, (iv) small fruit?

The Norfolk four-course rotation (turnips → barley → clover → wheat) introduced during the eighteenth century is still practised. The inclusion of sugar beet is the main modification.

331 Specialised Production and Land Use in Eastern England

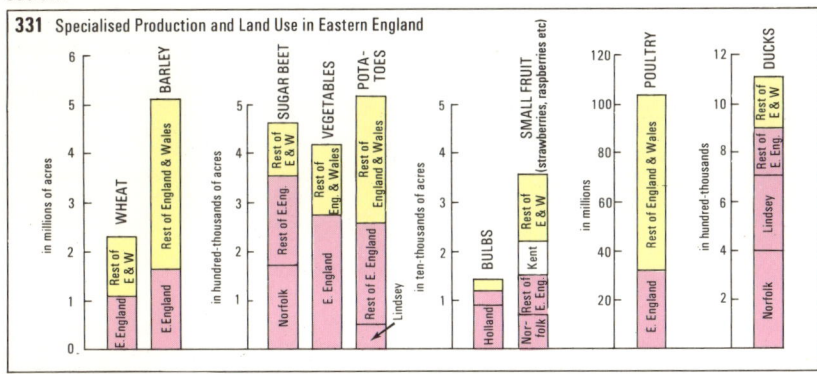

Key for Fig. 333:
- Silt Fen
- Peat Fen
- Land reclaimed from the sea since the 17th. century
- Areas which can be reclaimed in the future
- Mainly boulder clay

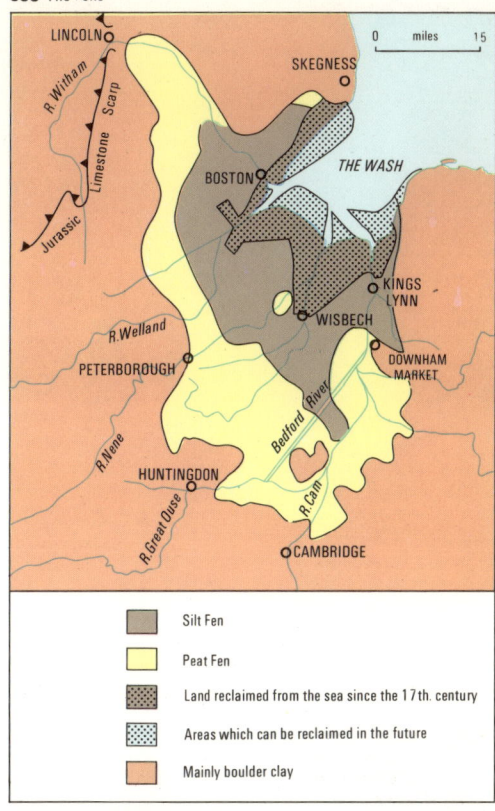

334(i) The Norfolk Broads

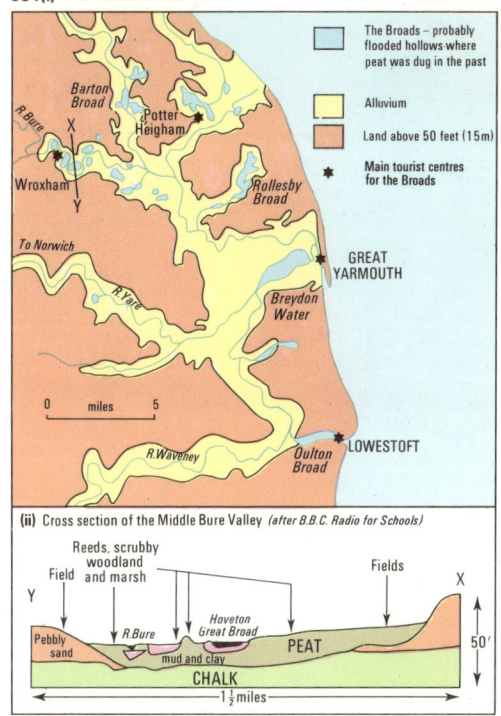

Figs. 332 and 333 The Fens

332a What are the two main types of land use west of Downham Market?

332b Explain (i) the regular field pattern (ii) the location of pastureland.

333a Describe the position of Downham Market.

333b Which counties or parts of counties occupy the Fens? (See Fig. 329.)

333c (i) Has the silt or peat fen been more recently reclaimed?
(ii) Are fen soils fertile? (See Fig. 324.)

333d What crops are grown in the Fens?

333e List (i) the main rivers flowing into the Wash, (ii) the main towns through which each flows.

333f How long ago were Boston and Kings Lynn seaports rather than river ports?

333g Explain the common situations of Lincoln, Peterborough, Huntingdon and Cambridge.

332 Land Use near Downham Market

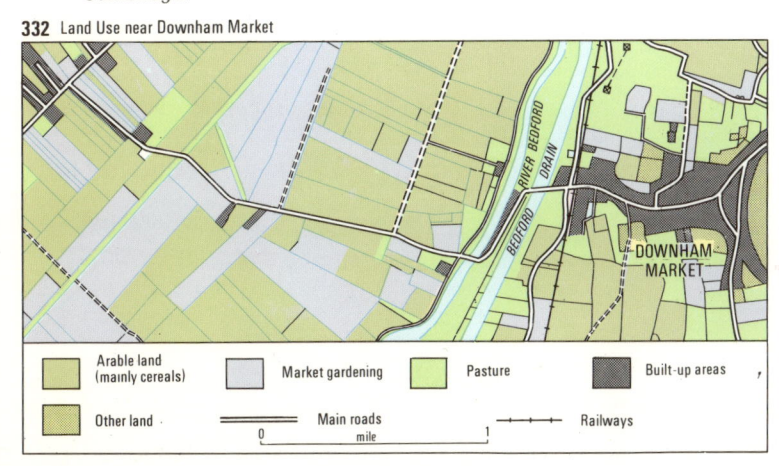

Key:
- Arable land (mainly cereals)
- Market gardening
- Pasture
- Built-up areas
- Other land
- Main roads
- Railways

(ii) Cross section of the Middle Bure Valley (after B.B.C. Radio for Schools)

Fig. 334 The Norfolk Broads

334a (i) Name three of the rivers that flow through the Norfolk Broads.
(ii) Where does each enter the sea?

334b Find out about, and explain, the formation of the Norfolk Broads.

334c (i) What is the main use of the Broads today?
(ii) Name three centres.

EASTERN ENGLAND Industry, Settlement

335(i) Brick-making in Eastern England

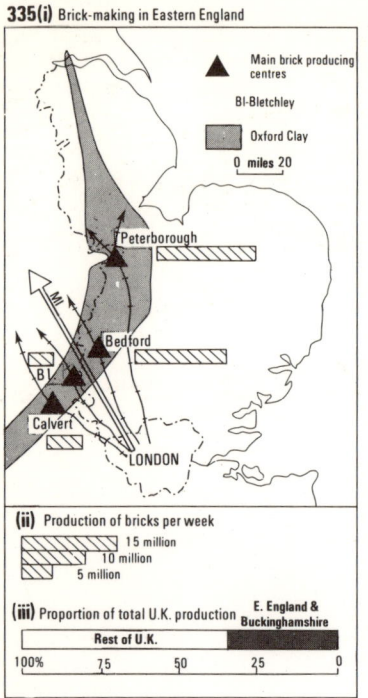

336(i) Manufacturing Industries

Industry in Eastern England

336(ii) Agricultural Processing and Natural Gas

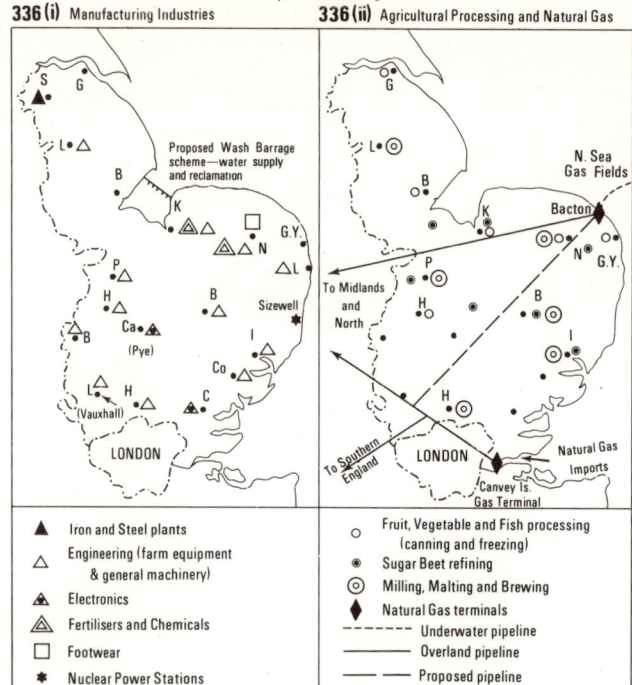

339 Sizes of Main Towns in E. England

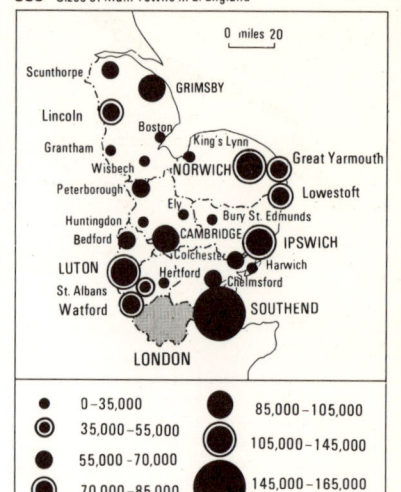

- 0–35,000
- 35,000–55,000
- 55,000–70,000
- 70,000–85,000
- 85,000–105,000
- 105,000–145,000
- 145,000–165,000

Key (335): Main brick producing centres; Bl–Bletchley; Oxford Clay

(ii) Production of bricks per week: 15 million / 10 million / 5 million

(iii) Proportion of total U.K. production — E. England & Buckinghamshire / Rest of U.K.

Key (336(i)): Iron and Steel plants; Engineering (farm equipment & general machinery); Electronics; Fertilisers and Chemicals; Footwear; Nuclear Power Stations

Key (336(ii)): Fruit, Vegetable and Fish processing (canning and freezing); Sugar Beet refining; Milling, Malting and Brewing; Natural Gas terminals; Underwater pipeline; Overland pipeline; Proposed pipeline

Figs. 335 and 336 Industry

335a Name the two main brick-making centres in Eastern England.

335b What factors favoured the industry in these locations?

335c What proportion of British brick production comes from this part of the Oxford Clay Belt?

Figs. 337 and 338 Ports

337a (i) Name the three main fishing ports shown.
(ii) Which is now the most important?

Yarmouth kippers and bloaters are still famous, but cannot compete with those from the Humber and Scotland.

337b (i) Which port in Eastern England handles the greatest annual tonnage of trade?
(ii) On what river is it situated?
(iii) How far is it from the sea?

337c (i) Which is the most important exporting port?
(ii) How far is it from London?

336a What is the most common manufacturing industry in Eastern England? Name one centre.

336b Name three other manufacturing industries and their centres.

336c (i) Name the natural gas terminals shown.
(ii) How do their supplies of natural gas differ?
(iii) Name six continental ports with which Eastern England trades.

On short sea journeys it is profitable to transport loaded lorries daily, thus avoiding loading and unloading ('roll-on/roll-off'). For the same reason the container is now much used.

338a What is a 'container'?

338b How do the 'roll-on/roll-off' and container facilities at Harwich compare with those at London?

338c Harwich is also an important passenger port. How did it compare with Dover in 1970?

Figs. 339, 340 and 341 Towns

339a Name the three largest towns in Eastern England.

339b What is Southend's main function?

340a (i) On what river did Norwich grow up?
(ii) What indications are there today of its original site?

Norwich, a classic example of a route centre, is the regional capital and market of East Anglia.

340b Name some of the towns to which it is linked by road.

340c What manufacturing industries occur in Norwich?

341a Lincoln is a gap town. Explain this term.

341b What effect has a gap on the local pattern of settlement and communication?

340 Site of Norwich

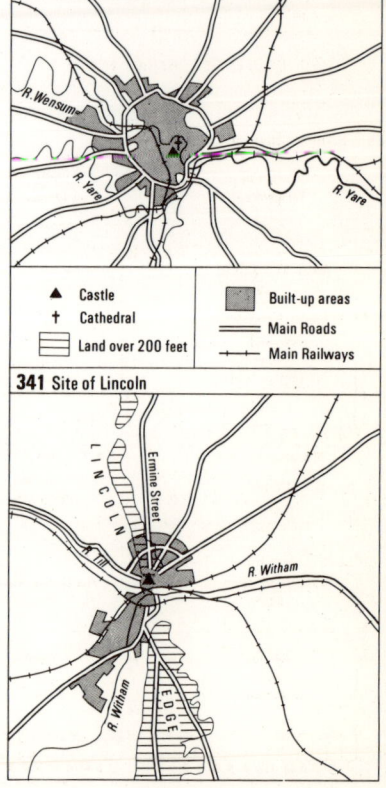

Key: Castle; Cathedral; Built-up areas; Main Roads; Land over 200 feet; Main Railways

341 Site of Lincoln

337 Eastern England—Fishing and Seaports

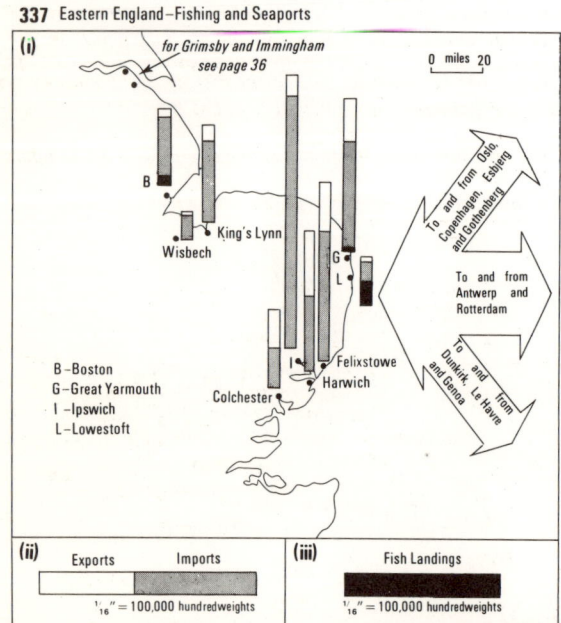

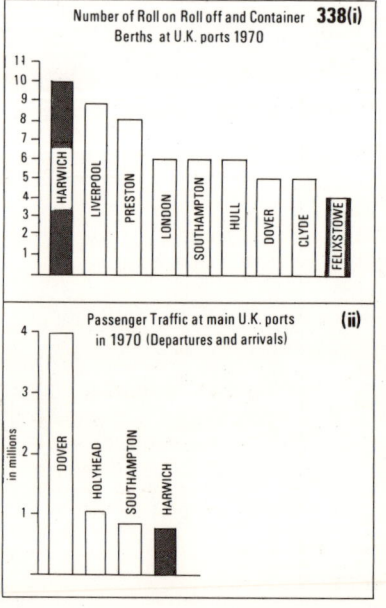

(i) for Grimsby and Immingham see page 36

B — Boston
G — Great Yarmouth
I — Ipswich
L — Lowestoft

To and from Oslo, Copenhagen, Esbjerg and Gothenburg
To and from Antwerp and Rotterdam
To and from Dunkirk, Le Havre and Genoa

338(i) Number of Roll on Roll off and Container Berths at U.K. ports 1970

HARWICH, LIVERPOOL, PRESTON, LONDON, SOUTHAMPTON, HULL, DOVER, CLYDE, FELIXSTOWE

(ii) Passenger Traffic at main U.K. ports in 1970 (Departures and arrivals) in millions

DOVER, HOLYHEAD, SOUTHAMPTON, HARWICH

(ii) Exports / Imports 1/16" = 100,000 hundredweights
(iii) Fish Landings 1/16" = 100,000 hundredweights

SOUTHERN ENGLAND Physical

342 Southern England—Geology

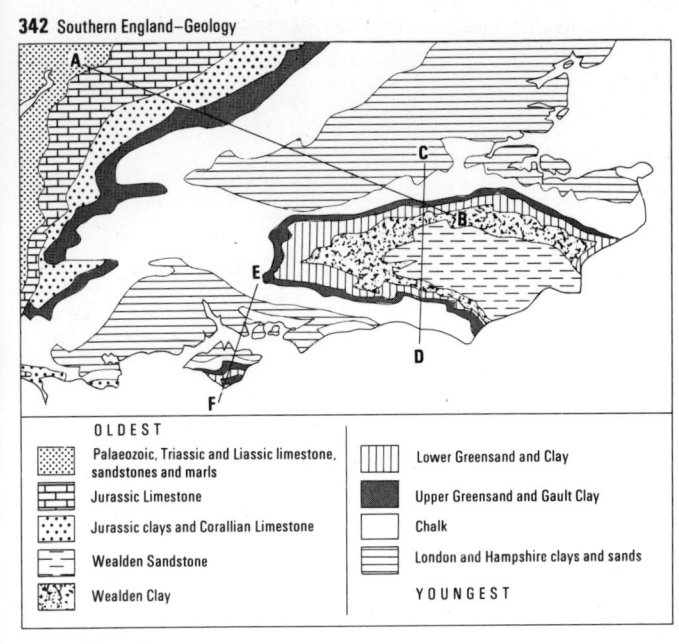

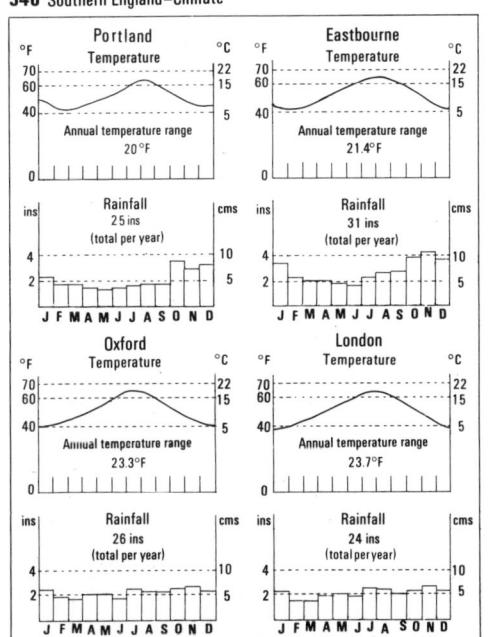

344 The North Downs Escarpment

Legend (OLDEST to YOUNGEST):
- Palaeozoic, Triassic and Liassic limestone, sandstones and marls
- Jurassic Limestone
- Jurassic clays and Corallian Limestone
- Wealden Sandstone
- Wealden Clay
- Lower Greensand and Clay
- Upper Greensand and Gault Clay
- Chalk
- London and Hampshire clays and sands

Fig. 342 Geology

342a (i) Name the youngest rocks in Southern England
 (ii) Where do they occur?
 (iii) Do they form highland or lowland?

343 Southern England—Relief and Drainage

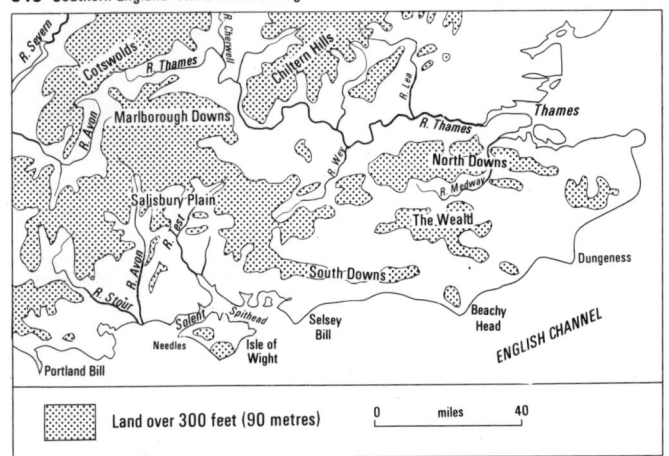

Land over 300 feet (90 metres) — 0 miles 40

Figs. 343, 344 and 345 Relief and drainage

343a (i) Does chalk form highland or lowland? Name four examples.
 (ii) Name two other rocks forming highland in Southern England.

343c What rocks most commonly form lowland?

343d Name (i) two rivers that rise in the Cotswolds, (ii) two tributaries of the river Thames, (iii) three headlands on the south coast (include a chalk outcrop).

344a What is an escarpment?

For lines of cross-section, see Fig. 342.

345a Name (i) three escarpments in Southern England, (ii) the rock they are made of.

345b What is (i) an anticline, (ii) a syncline? (iii) Give one example of each in Southern England.

345 Southern England—diagrammatic Cross-sections

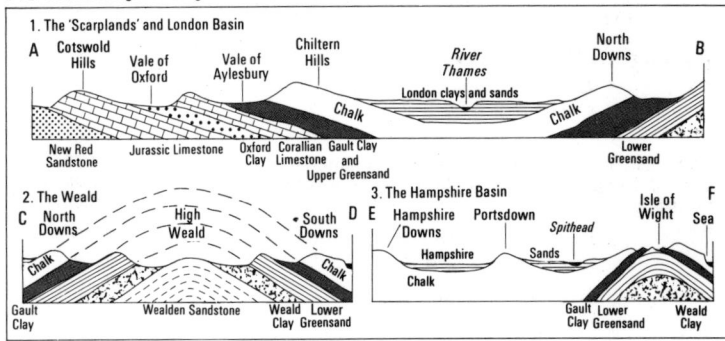

1. The 'Scarplands' and London Basin — Cotswold Hills, Vale of Oxford, Vale of Aylesbury, Chiltern Hills, River Thames, London clays and sands, North Downs; New Red Sandstone, Jurassic Limestone, Oxford Clay, Corallian Limestone and Upper Greensand, Gault Clay, Chalk, Lower Greensand
2. The Weald — North Downs, High Weald, South Downs; Chalk, Gault Clay, Wealden Sandstone, Weald Clay, Lower Greensand
3. The Hampshire Basin — Hampshire Downs, Portsdown, Spithead, Isle of Wight, Sea; Hampshire Sands, Chalk, Gault Clay, Lower Greensand, Weald Clay

Fig. 346 Climate

346a Find the position (see Fig. 347) of the towns named in Fig. 346.

346b (i) Which town experiences the warmest winter temperature?
 (ii) What happens to winter temperatures as one goes east and north?
 (iii) Why does Eastbourne have warmer winter temperatures than London?

346c What is the average July temperature at Eastbourne?

346d (i) Explain 'temperature range'.
 (ii) Which places experience the smallest and greatest temperature range?
 (iii) Explain why.

346e Which town experiences least rainfall, and why?

346 Southern England—Climate

Portland — Annual temperature range 20°F — Rainfall 25 ins (total per year)
Eastbourne — Annual temperature range 21.4°F — Rainfall 31 ins (total per year)
Oxford — Annual temperature range 23.3°F — Rainfall 26 ins (total per year)
London — Annual temperature range 23.7°F — Rainfall 24 ins (total per year)

347 Southern England—Counties

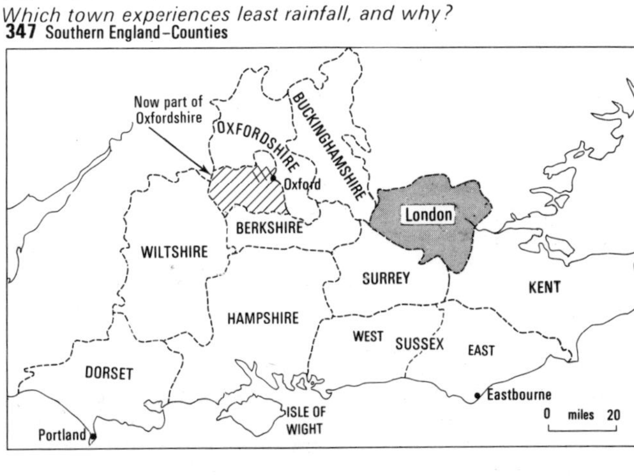

Fig. 347 Counties

347a Which counties of Southern England are (i) the driest, (ii) the mildest, (iii) the warmest, in summer?

SOUTHERN ENGLAND   Agriculture

348 Agricultural Land Use in Southern England (1)

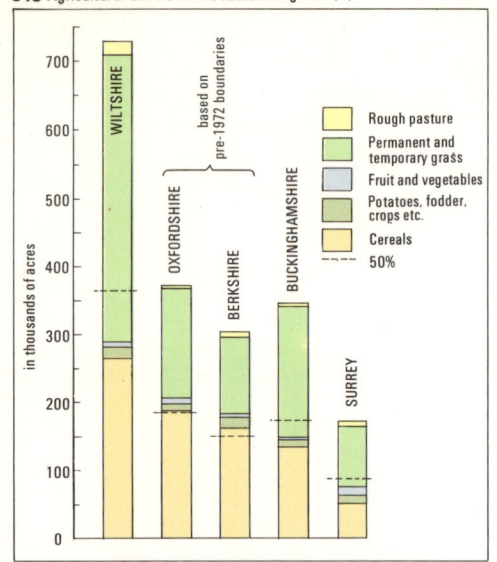

349 Agricultural Land Use in Southern England (2)
Key as in figure 348

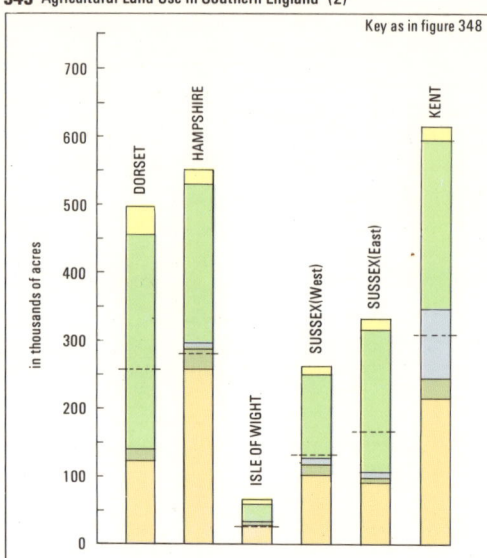

350 Specialised Agricultural Land Use

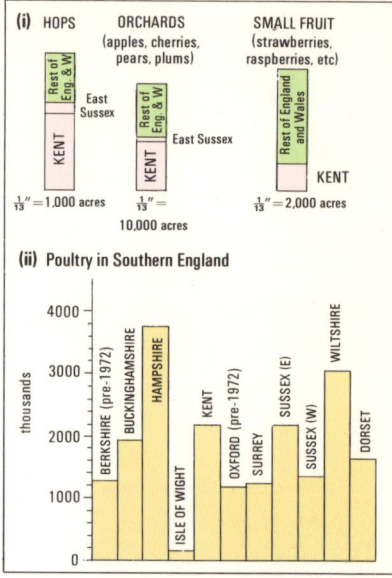

(i) HOPS — ORCHARDS (apples, cherries, pears, plums) — SMALL FRUIT (strawberries, raspberries, etc)
- KENT / Rest of Eng. & W / East Sussex
- $\frac{1}{13}''$ = 1,000 acres
- $\frac{1}{13}''$ = 10,000 acres
- $\frac{1}{13}''$ = 2,000 acres

(ii) Poultry in Southern England

Figs. 348, 349 and 350 Land use

- **348/350a** Choose one of the following words to describe the agriculture of Southern England: pastoral, mixed, arable (see also Fig. 249).
- **348/350b** Which county has the highest proportion of (i) grassland, (ii) arable land?
- **348/350c** What have the positions of each of these counties in common?
- **348/350d** Which is the most important county for fruit and vegetables?
- **348/350e** Explain why Kent has a large proportion of arable land.
- **348/350f** (i) What are the main fruits grown in Kent?
  (ii) Why are hops grown?
- **348/350g** What proportion has Kent of the total acreage under orchards in England and Wales?
- **348/350h** (i) What are the main poultry-producing counties in Southern England?
  (ii) Suggest their main market.

Fig. 352 Transect across Northern Weald

The great variety of farming in Southern England results from the variety of physical factors (geology, relief, soil, climate) and the changing market demands the farmer has tried to meet down the centuries.

- **352a** Describe how agriculture in the Plain of North Kent differs from that in the Vale of Kent.

351 Livestock in Southern England

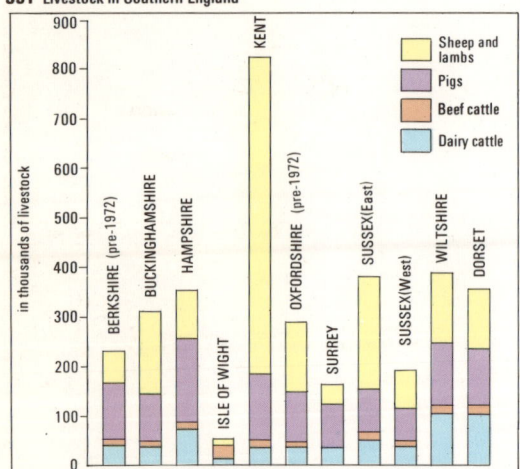

352 Transect across the Northern Weald

	Plain of North Kent	Chalk Dip Slope	Chalk Scarp Slope	Clay Vale	Greensand Ridge	Clay Vale	High Weald
Agriculture	Fruit growing and market gardening for London market. Orchard fruits (apples, cherries, pears, plums) and vegetables.	Cereals and fodder crops. Orchards on lower slopes	Woodland and permanent pasture for sheep grazing	Dairying on permanent pasture.	Woodland heath and rough pasture	Dairying and cultivation on improved soils and river terraces	Much forest and woodland e.g. Ashdown Forest. Some dairying and beef farming
Soil	Clays Fertile loams	Clay—with flints soil (in contrast to South Downs)	Thin dry soil	Heavy clay soils	Thin infertile sandy soil	Heavy wet clay soils in natural state	Mainly poor sandy soils

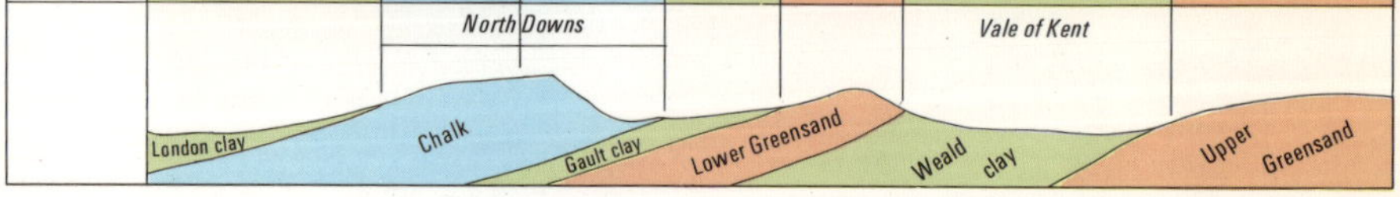

North Downs — Vale of Kent
London clay — Chalk — Gault clay — Lower Greensand — Weald clay — Upper Greensand

Fig. 351 Livestock

- **351a** (i) In what parts of Southern England is dairying most important?
  (ii) Explain why (think of climate and markets).
- **351b** (i) In which two counties are most sheep found?
  (ii) Explain why.
- **351c** Romney Marsh and the Isle of Pevensey are important sheep-rearing areas. In what counties do they occur?

Fig. 353 The Channel Islands

- **353a** What islands make up the Channel Islands?
- **353b** (i) Which British ports operate regular sea services to the islands?
  (ii) How far is the sea journey? (See Fig. 357.)
  (iii) What other transport links have the Channel Islands with Britain?

Agriculture and tourism are the mainstays of the Channel Islands' economy.

- **353d** (i) Describe the main forms of agriculture.
  (ii) What factors favour tourism? (Think of climate, continental atmosphere, customs duties, purchase tax, etc.)

353 The Channel Islands

(i) From Weymouth and Southampton — Alderney — Guernsey — Sark — Jersey — Cherbourg — Cotentin Peninsula (mainland France) — 49.5°N
Attached to English Crown since 1066 BUT unaffected by British law and custom duties

(ii) JERSEY — Dairying — St. Aubin — St. Hellier — Main outdoor tomato growing area

(iii) GUERNSEY — Early vegetables, flowers and fruit grown in glass-houses — St. Sampson — St. Peter Port — Dairying with some arable

- Igneous rocks
- Very old rocks (often metamorphosed)
- Land over 200 feet (60m)
- Land below 200 feet
- A Airport

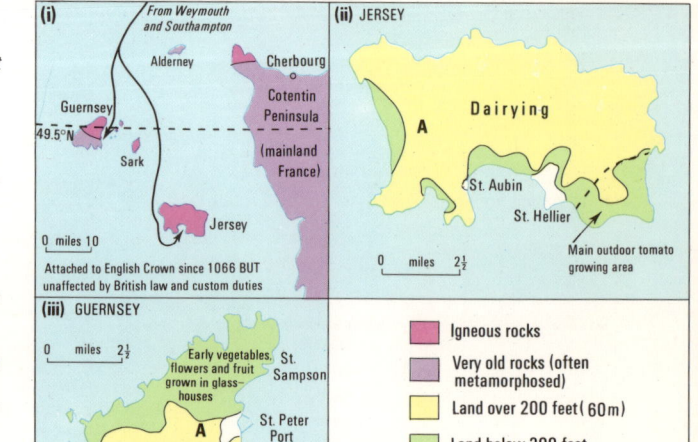

SOUTHERN ENGLAND Ports and Coastal Resorts

354 Southampton: Docks, Industry and Built-up areas

355 Tidal range at selected British ports

356 Southampton: Port traffic

- in millions of tons: 0 2 4 6 8 10 12 14 16 18
- (i) IMPORTS — Petroleum and Petroleum products, Man. Goods, Foodstuffs, Raw materials
- (ii) EXPORTS — Petroleum and Petroleum products, Man. goods, Foodstuffs and Raw materials
- (iii) Imports of Foodstuffs — Prpn of fruit and veg / Main sources of fruit and veg: Others, West Indies, Mediterranean Countries Iberia and France, Southern Africa
- (iv) Main sources of imported petroleum — Mediterr., Pers. Gulf, Others
- (v) Main destinations of 'exported' petroleum — Other British ports
- (vi) Passengers arriving and departing at British Ports—1968: Dover, Holyhead (mainly long distance), Southampton, Harwich, Others. $\frac{1}{10}'' = \frac{1}{2}$ million passengers

Built-up area / Forest and woodland / Roads / Railways

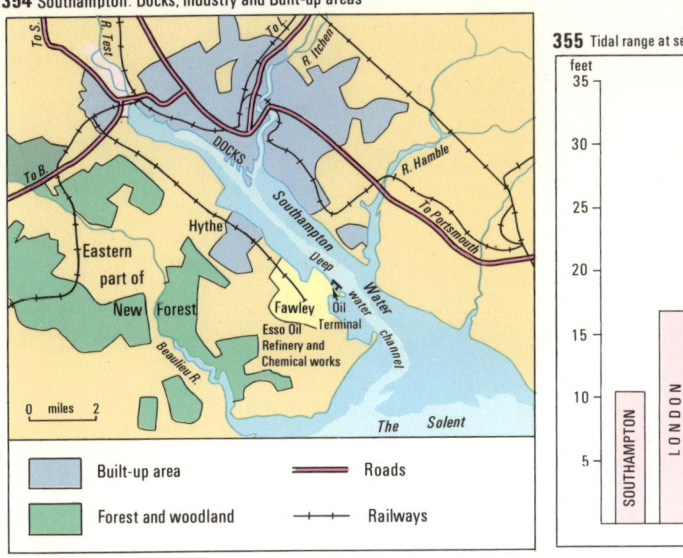

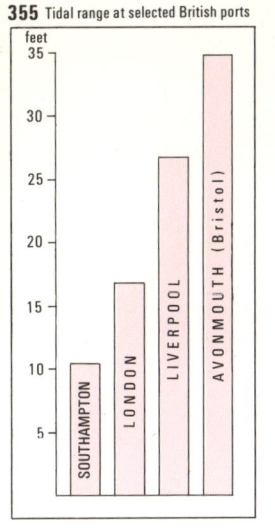

Fig. 354 Southampton

Southampton grew up between two rivers (name them), which become Southampton Water. It is the regional centre of central Southern England. Its modern development began in 1836 with the arrival of the railway from London, and the formation of a dock company. It is still essentially an outlet for London.

354a On which bank of Southampton Water has Southampton grown up? (Suggest why.)

354b Name the channels that link Southampton Water with the English Channel (Fig. 359).

Fig. 355 Tidal ranges

355a (i) Explain 'tidal range'.
(ii) Has Southampton a high or low tidal range?
(iii) In what way is tidal range important in port construction?

355b (i) Are Southampton's docks open or enclosed?
(ii) Explain why.

357(i) Southern England: Ferry ports

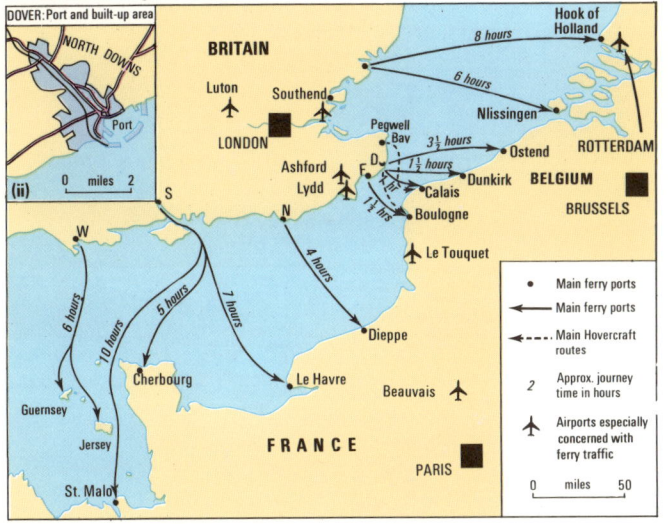

358 Passengers arriving and departing at Dover and Folkestone

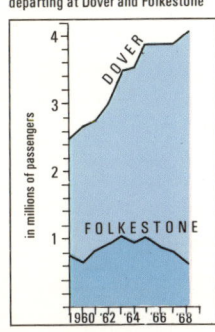

Fig. 357 and 358 Ferry ports

357a (i) List the main ferry ports on the south coast of England.
(ii) Which is the most important? (See Fig. 356.)
(iii) Explain why.

357b Name the British airports engaged in cross-channel ferrying.

357c What is meant by 'Chunnel'?

358a Describe the relative variation in traffic volume at Folkestone and Dover since 1959.

Fig. 360 Portsmouth

360a Portsmouth is a naval base and dockyard. Describe (i) its site, (ii) the factors that favoured its development as a naval base.

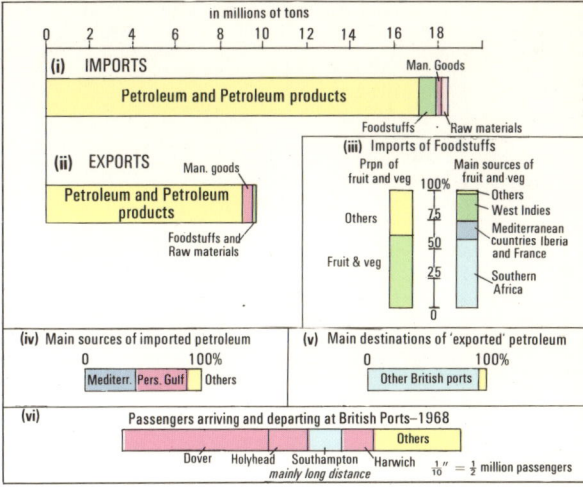

Fig. 356 Southampton's trade

356a (i) What is Southampton's main import?
(ii) Where is it obtained?
(iii) Where is it refined?

356b (i) What are the main types of foodstuffs imported?
(ii) Where are they obtained?

356c (i) Compare the passenger traffic of Dover and Southampton in terms of numbers and type.
(ii) What advantage has Southampton over London as a world passenger port?

359 Southern England: Coastal and inland tourist centres

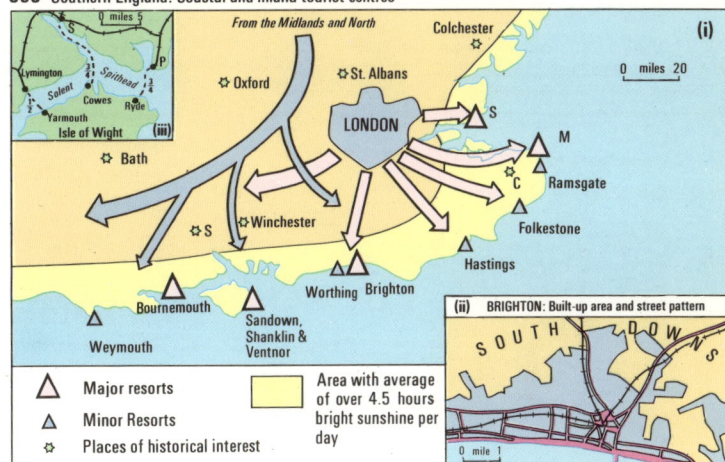

(ii) BRIGHTON: Built-up area and street pattern

Major resorts / Minor Resorts / Places of historical interest / Area with average of over 4.5 hours bright sunshine per day

Fig. 359 Tourist centres

359a (i) What are the five major coastal resorts in Southern England?
(ii) What advantages has this region? (Think of summer temperatures, rainfall, sunshine hours, proximity to large urban centres.)

359b (i) Name the I.O.W. ferry routes by terminals.
(ii) Figures are crossing times in hours. Which is the quickest route?

359c Describe the street pattern of Brighton.

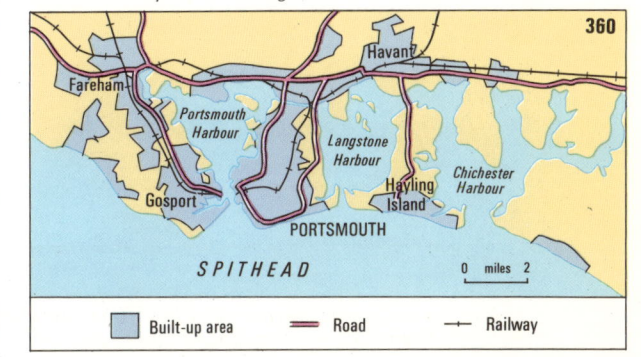

360 Portsmouth — Fareham, Havant, Portsmouth Harbour, Langstone Harbour, Gosport, Hayling Island, Chichester Harbour, PORTSMOUTH, SPITHEAD

Built-up area / Road / Railway

SOUTHERN ENGLAND Industry and Towns

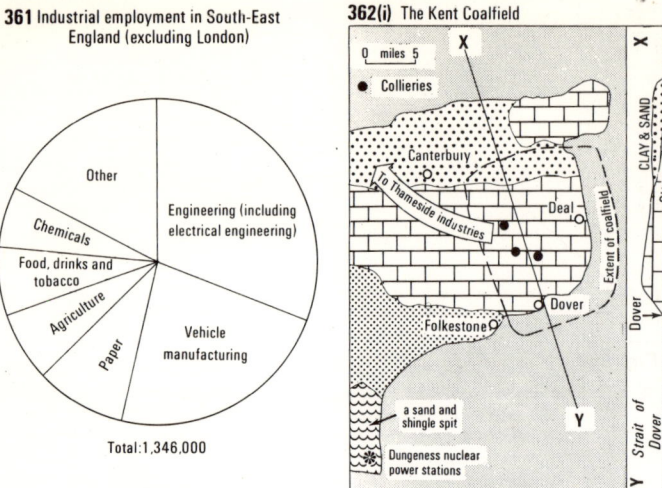

Fig. 361 Industry

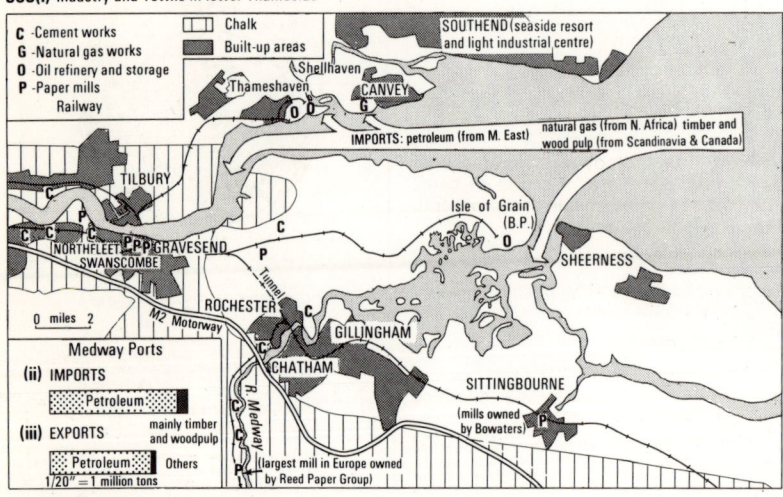

361a (i) List the main industries in South-East England.
(ii) Under which of the following categories does each of them come: heavy basic, light assembly, consumer, primary (e.g. fishing, mining or quarrying).

Fig. 362 The Kent coalfield

Discovered in the 1880s, this is one of Britain's more recent coalfields.

362a (i) Is it exposed or concealed?
(ii) At what depth are the coal measures?
(iii) How many collieries are active today?
(iv) What is the annual coal output?

362b Suggest why little industrialisation has taken place on this coalfield.

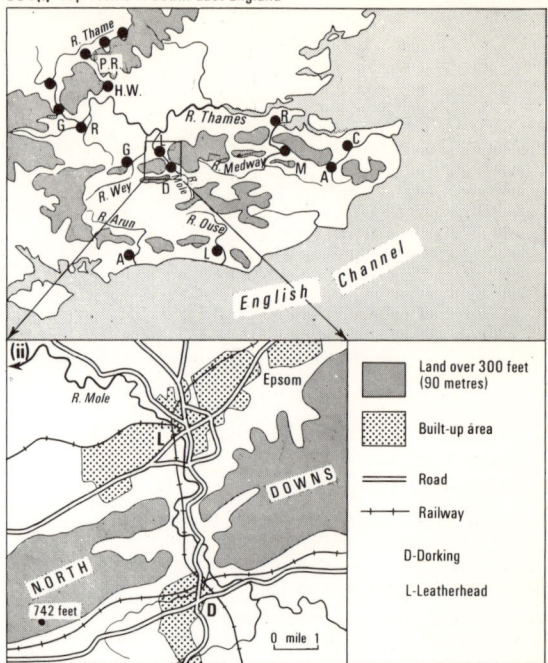

Fig. 364 Gap towns

364a (i) Name the three lines of hills in Southern England through which gaps have been eroded.
(ii) Give one gap example for each (by naming the town(s) and river).

Fig. 363 Industry—towns on lower Thameside

363a (i) What is the main industry in the Medway valley?
(ii) Chalk and mud are important for this industry. Where are they obtained?
(iii) What fuel is used? (See Fig. 362.)

The paper-making industry is also important.

363b Name (i) four paper-making centres, (ii) two firms with large mills.

363c What factors favour paper-making here, (think of raw materials and fuel). Water purity is controlled artificially.

363d (i) What is the Medway ports' major import?
(ii) Name three oil refineries on Thameside.
(iii) What other fuel is imported?

Figs. 365 and 366 Oxford and Salisbury

365/366a Compare (i) the site, (ii) the situation, (iii) the function of Oxford and Salisbury.

These two towns are representative of many in Southern England.

365 Oxford

GROWTH
1. Gravel terrace provided bridging point over River Cherwell
2. Upstream limit of commercial navigation of Thames
3. Occupies gap through minor limestone escarpment

FUNCTION
1. Market town
2. University town (established before 1200)
3. Tourist centre
4. Industrial centre—
  (a) Car assembly (British Leyland)
  (b) Car bodies (Pressed Steel Co.)
  (c) Printing

366 Salisbury

GROWTH
1. Original site at Old Sarum (a defensive site) abandoned in thirteenth century
2. New site in valley-bridging point of Rivers Avon and Bourne
3. Situated at point where river valleys converge

FUNCTION
1. Market centre (linked to its situation as a route centre)
2. Ecclesiastical centre (cathedral)
3. Tourist centre
4. Industrial centre (mainly light engineering)

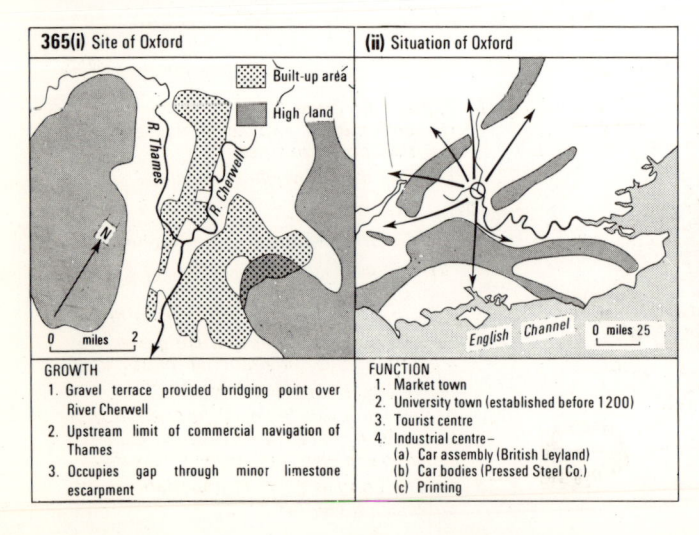

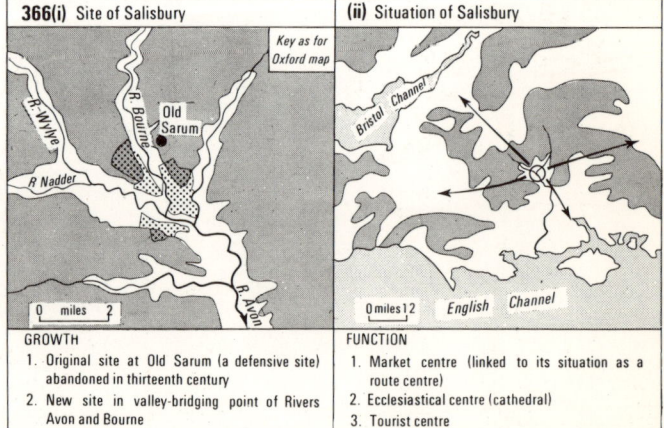

ENGLAND London

367(i) The situation of London

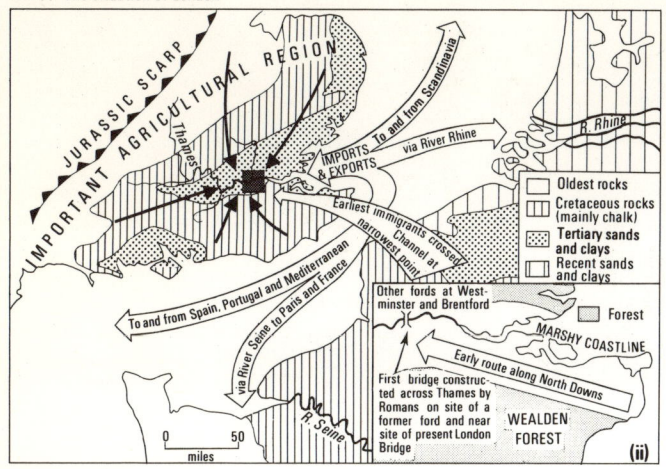

Fig. 367 Situation

The route along the North Downs directed the earliest immigrants to the Thames by present-day central London; here they found convenient fording and bridging places.

367a What channel crossing-point did the earliest immigrants take to Britain?

368 The site of London

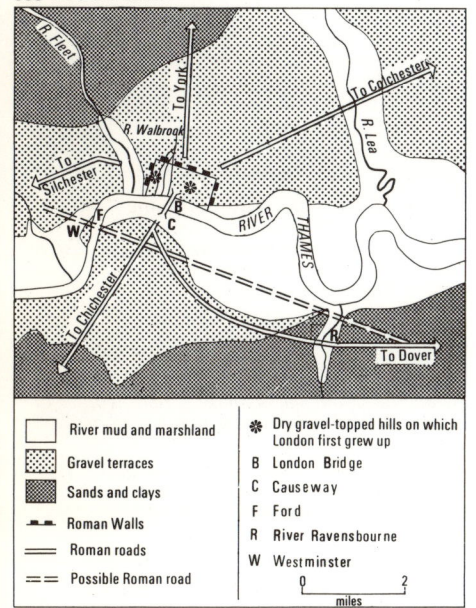

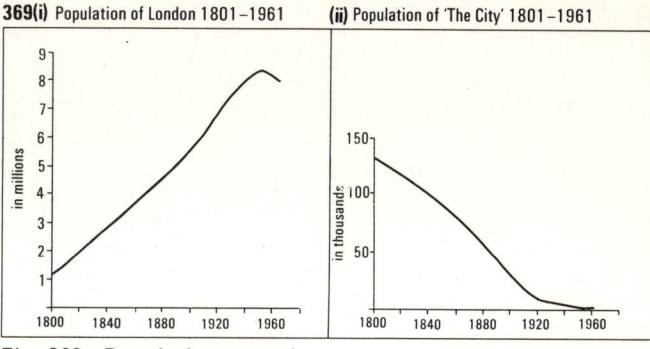

Key: River mud and marshland; Gravel terraces; Sands and clays; Roman Walls; Roman roads; Possible Roman road; Dry gravel-topped hills on which London first grew up; B London Bridge; C Causeway; F Ford; R River Ravensbourne; W Westminster

369(i) Population of London 1801–1961   **(ii)** Population of 'The City' 1801–1961

Fig. 369 Population growth

369a What has happened to the population of London since 1800? The apparent decline since 1950 is due to growth taking place outside London's boundary, where people are being rehoused from the overcrowded central areas.

369b Explain the decline of population in the 'City'.

367b Describe (i) London's situation in agricultural England in Roman times, (ii) the river Thames as a route-way, (iii) London's early trading partners in Europe, (iv) the London Basin's geology.

Fig. 368 Site

368a What advantages persuaded the Romans to build the first London Bridge where they did? (Consider width of river mud and marsh, dry site for settlement away from flood plain of river, defence.)

368b The Roman 'Londinium' was one mile square.
 (i) On which bank of the Thames was it established?
 (ii) What natural defence existed in the west and south?

368c With which towns was 'Londinium' linked by road?

The Roman town corresponds to the present-day 'City' of London. After the Roman withdrawal in the fifth century little is known of London until the Danes restored its role as a centre of commerce. A new period of expansion and prosperity followed the Norman conquest.

Fig. 370 Areal growth

370a Describe the growth of London between 1850 and 1966.

Fig. 371 Employment

371a (i) What proportion of the total working population of London is engaged in manufacturing?
 (ii) What are the main manufacturing industries?

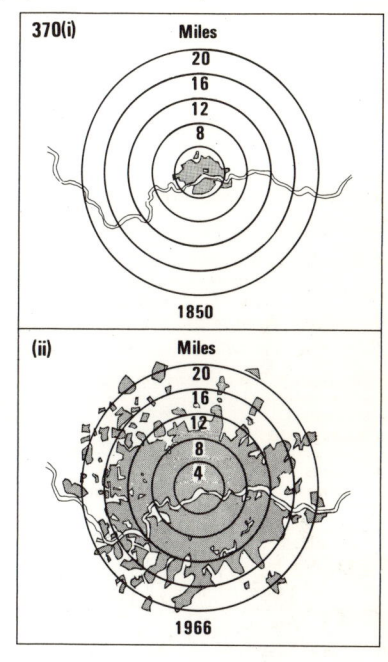

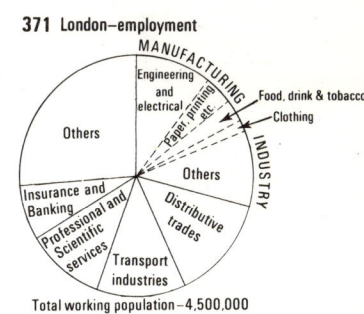

370(i) 1850   **(ii)** 1966

371 London—employment

Total working population – 4,500,000

371b Explain the terms: (i) distributive trades, (ii) professional and scientific services, (iii) transport industries.

371c Why are they so well represented?

Fig. 372 'The City'

This is the hub of British commerce and finance.

372a (i) What is a 'wholesale market'? Give three examples.
 (ii) In what type of goods do they deal?

372b An exchange is a place where goods are auctioned and international prices set. Name four examples in the 'City'.

372c Name the main 'financial houses' in the 'City'. Many banks, insurance companies, building societies, etc., have offices there.

372d Why is London such an important world commercial centre?

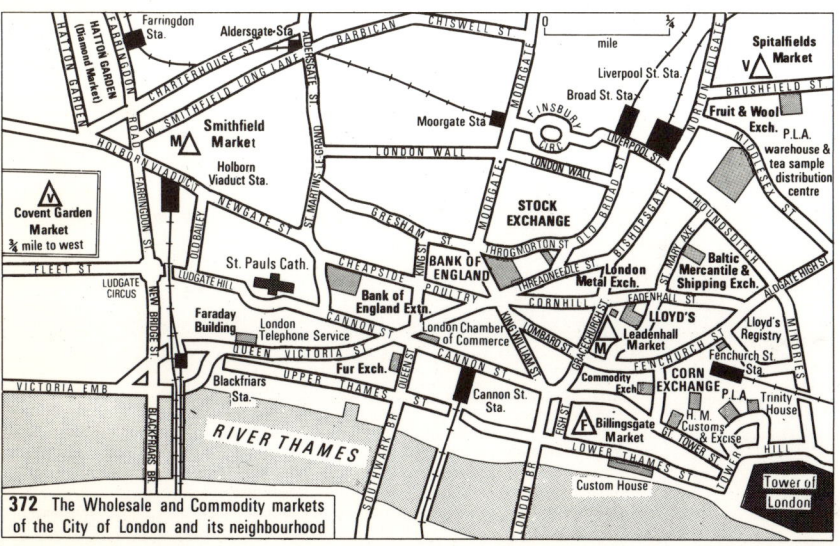

372 The Wholesale and Commodity markets of the City of London and its neighbourhood

ENGLAND London

373 Railway Termini in Central London

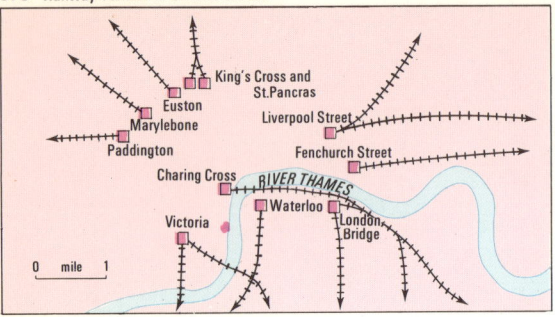

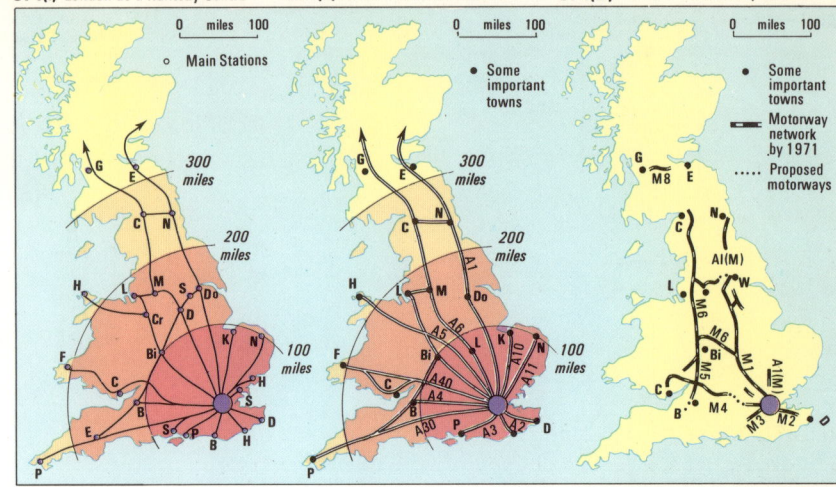

Figs. 373 and 374 Road and rail centre

373a (i) List London's railway termini under two headings: 'North' and 'South' of the river Thames.
(ii) Which termini are north of the river, but serve areas to the south?

374a Describe Britain's (i) railway pattern, (ii) road pattern, (iii) motorway pattern.

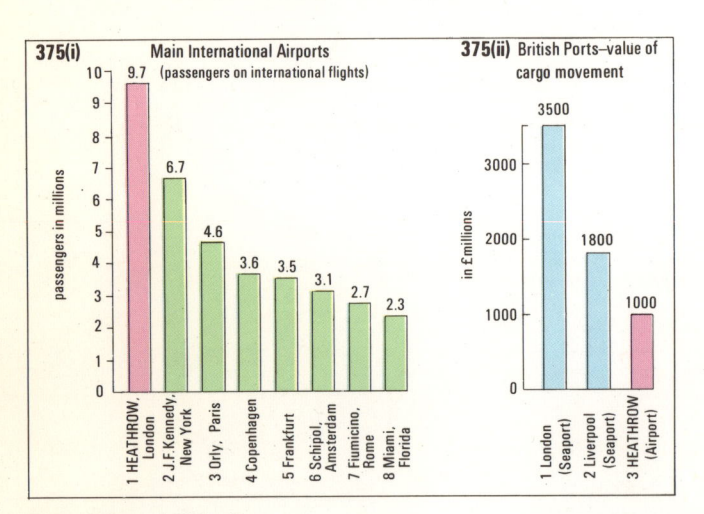

374(i) London as a Railway Centre

374(ii) London as a Road Centre

374(iii) London–a Motorway Centre?

375(i) Main International Airports (passengers on international flights)

1 HEATHROW, London — 9.7
2 J.F. Kennedy, New York — 6.7
3 Orly, Paris — 4.6
4 Copenhagen — 3.6
5 Frankfurt — 3.5
6 Schipol, Amsterdam — 3.1
7 Fiumicino, Rome — 2.7
8 Miami, Florida — 2.3

375(ii) British Ports–value of cargo movement (in £millions)

1 London (Seaport) — 3500
2 Liverpool (Seaport) — 1800
3 HEATHROW (Airport) — 1000

Figs. 375–378 Air Traffic Centre

375a Give figures to show how London ranks as an international air centre.
375b (i) Compare London's air- and seaports in terms of cargo traffic.
(ii) Would the difference be reduced or exaggerated if tonnage instead of value of cargo traffic were given?

376a Name at least one centre in each continent with which London has regular air contact.

377a (i) Name London's two main airports.
(ii) Which is the more important?
377b (i) Name the four sites considered for London's third airport.
(ii) What factors have to be considered when deciding on the site of a new airport?

378a Work out the average annual increase of passenger traffic at Heathrow during the 1960s.
378b If this rate of increase continues, how many passengers would use Heathrow in 2000 A.D.?

377 London's Airports

- London's main airports
- London's other airports
- Four alternative sites for London's new airport (1970)
- Motorway
- Greater London boundary

Thurleigh, Nuthampstead, Cublington, Luton, Stanstead, Southend, Foulness, Northolt, Heathrow, Gatwick

378 Heathrow-passenger traffic (1960-68)

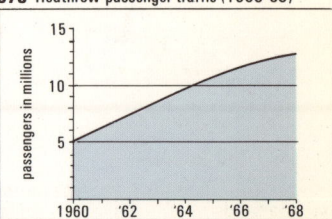

376 London-International Air Centre

Figs. 379 and 380 Port

London first developed as a port in Roman times, when the low-arched bridge prevented sea-going ships from proceeding further upstream (i.e. London became the effective head of navigation). Modern expansion can be dated from Elizabethan times.

379a (i) When and why were the first dock basins constructed?
(ii) Why are the Surrey Commercial Docks the only docks south of the Thames?
(iii) Which are the largest docks?
(iv) When were they constructed?

380a (i) How far downstream from the Royal Docks are Tilbury Docks (1886)?
(ii) What advantage have they over the other London Docks?
380b Name some of the modern cargo-handling methods in use at Tilbury.

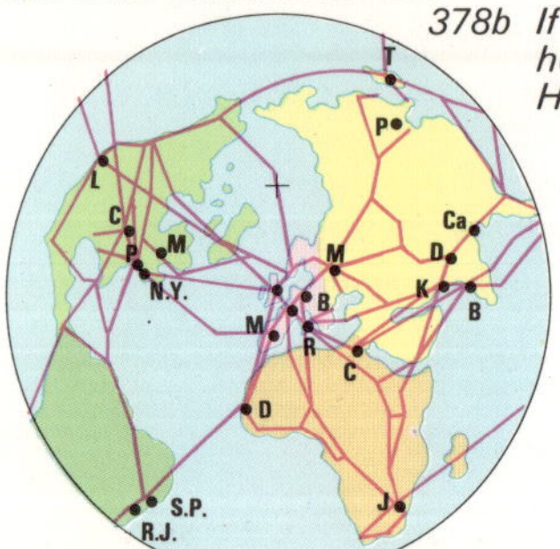

379 Development of the Port of London

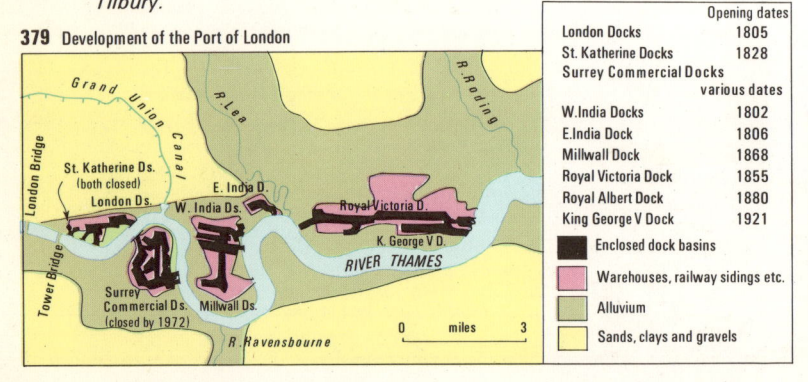

Opening dates	
London Docks	1805
St. Katherine Docks	1828
Surrey Commercial Docks	various dates
W. India Docks	1802
E. India Dock	1806
Millwall Dock	1868
Royal Victoria Dock	1855
Royal Albert Dock	1880
King George V Dock	1921

- Enclosed dock basins
- Warehouses, railway sidings etc.
- Alluvium
- Sands, clays and gravels

380 Tilbury Docks

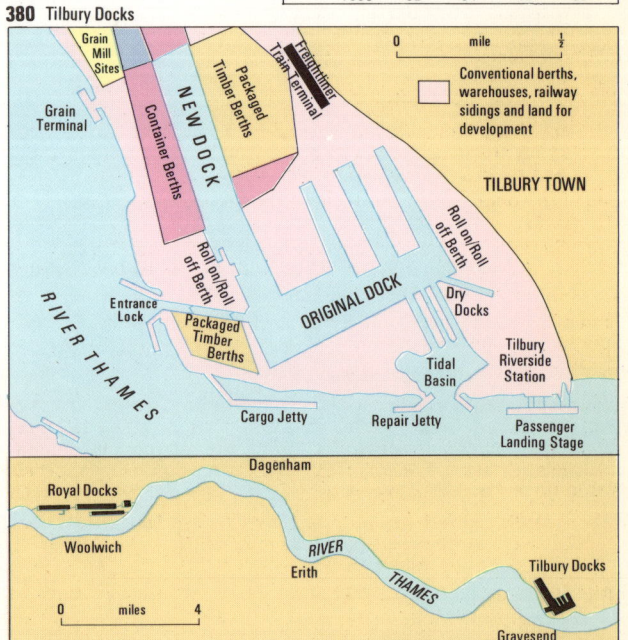

Conventional berths, warehouses, railway sidings and land for development

Grain Mill Sites, Grain Terminal, Freightliner Train Terminal, NEW DOCK, Packaged Timber Berths, Container Berths, Roll on/Roll off Berth, Packaged Timber Berths, Entrance Lock, ORIGINAL DOCK, Roll on/Roll off Berth, Dry Docks, TILBURY TOWN, Tidal Basin, Tilbury Riverside Station, Cargo Jetty, Repair Jetty, Passenger Landing Stage, Royal Docks, Dagenham, Woolwich, Erith, Tilbury Docks, Gravesend

ENGLAND London

381 Traffic of the Port of London
(i) Foreign Traffic (36.7 mill. tons)
(ii) Coastwise Traffic (19.6 mill. tons)

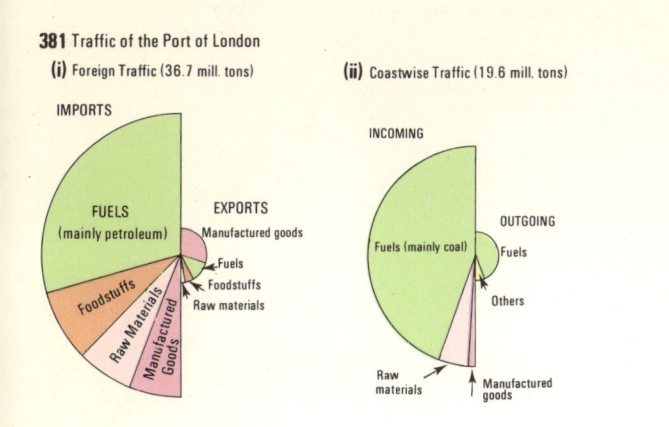

382 Proportion of traffic in selected commodities handled by Port of London
(i) Foreign Imports
(ii) Foreign Exports

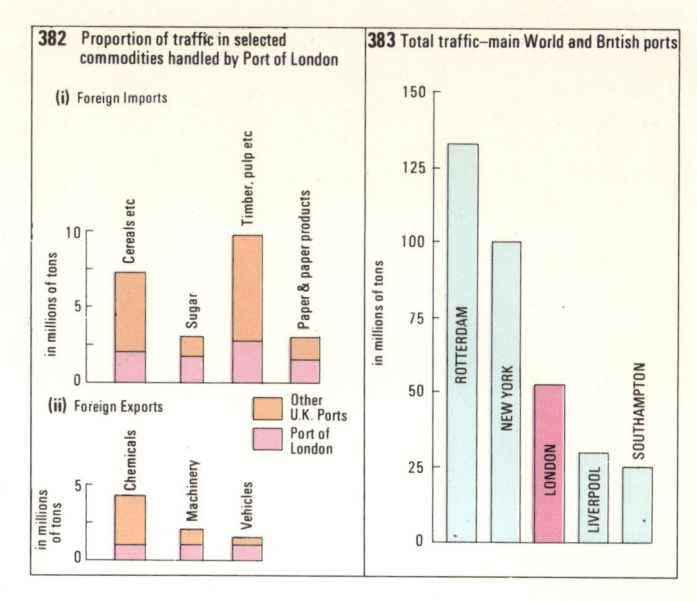

383 Total traffic – main World and British ports

Figs. 381, 382 and 383 Trade

A port that imports a large volume of foreign goods and then re-exports some of them is called an entrepôt. London is an entrepôt.

381a (i) What is the total annual traffic of the Port of London?
(ii) Are foreign imports or exports most important?

381b What proportion of total foreign imports and exports is made up of (i) fuels, (ii) manufactured goods?

381c (i) In what way do London's foreign and coastwise imports of fuel differ?
(ii) Where is most of the coastwise fuel obtained?

382a The Port of London specialises in particular traffics.
(i) Name two commodities imported into Britain mainly via London.
(ii) Explain why London dominates the export trade in vehicles and machinery.

383a How does London rank as a world port?

384 The main industrial areas of Greater London

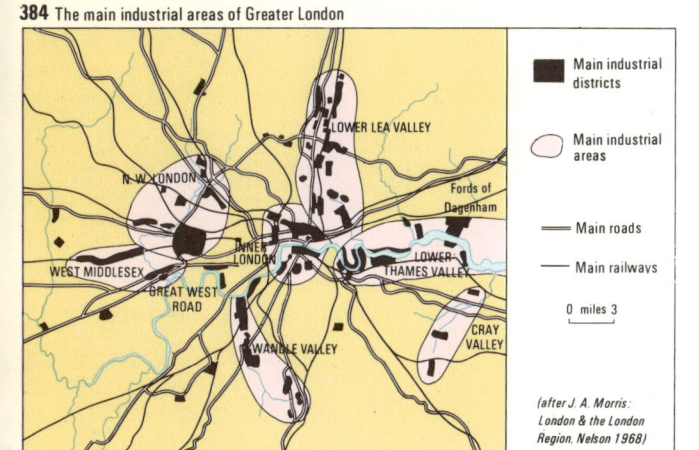

Fig. 384 Industries

London has always been an important industrial centre, although remote from coalfields or sources of raw material. London's industries include: clothing, cement, oil refining, jewellery, furniture, cosmetics, flour-milling, leather-working, television, gas production, records, metal-working, confectionery, vehicle assembly.

384a List each industry under one of the following headings: (i) Traditional Craft Industries (concentrated in Inner London), (ii) Bulky Industries (using 'imported' raw materials), (iii) Consumer Goods Industries (those relying on London's vast population for their labour supply and market).

384b List London's four main manufacturing industries. (See Fig. 371, p. 61.)

384c Suggest ways in which industry of the lower Thames Valley differs from that of North-West London (raw materials, distribution of finished goods, etc.).

London is also the capital of Britain and the Commonwealth, and an important cultural, tourist, shopping and entertainment centre.

Fig. 386 New and expanding towns

Redevelopment of the old congested central areas, and continual movement into London, have resulted in overspill (or 'surplus') population. Despite the common use of blocks of flats, the new population density in redeveloped areas does not match the original density. Hence people have to be re-housed outside London.

386a (i) Distinguish two ways in which London's overspill population is being accommodated.
(ii) Give two examples of each.

386b Name one method being used to try and check the physical growth of London.

385 London's water supply
(iii) The artesian basin under London
(i) (map showing Enfield, Chingford, Walthamstow, Reservoirs, Staines, Sunbury, Walton, 'West End', R. Thames, R. Lea)
(ii) Source of water distributed by Metropolitan Water Board — River Thames 70%, R. Lea 15%, Wells 15%

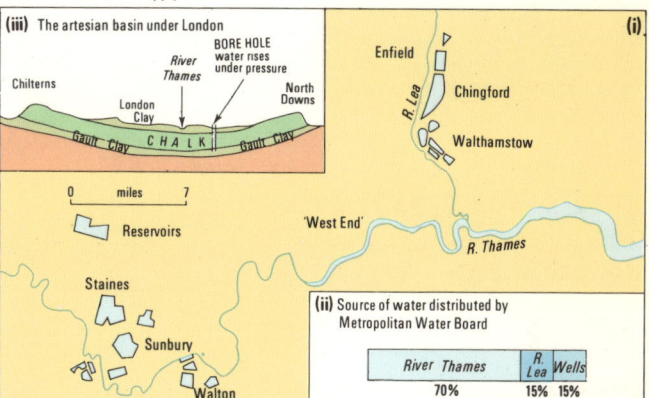

Figs. 385 Water supply

385a What are the three main sources of London's water supply?

386 New and expanding towns in South-East England
(ii) Plan of Basildon New Town

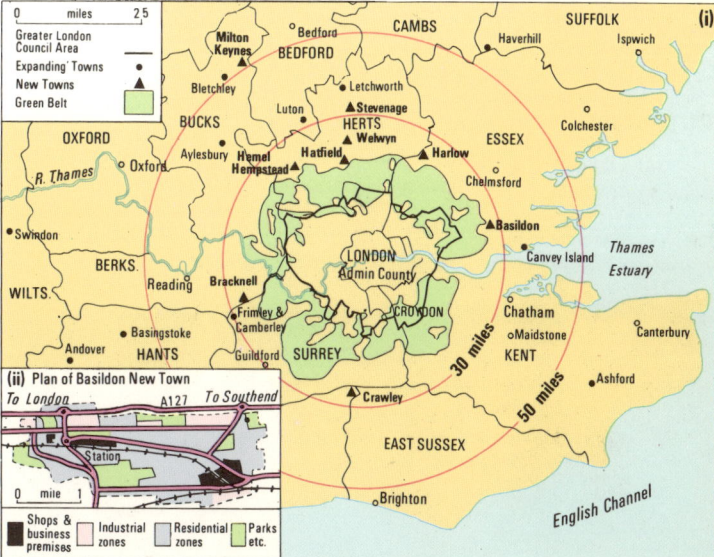

BRITAIN AND EUROPE   Interdependence

387 Western Europe-Economic groupings

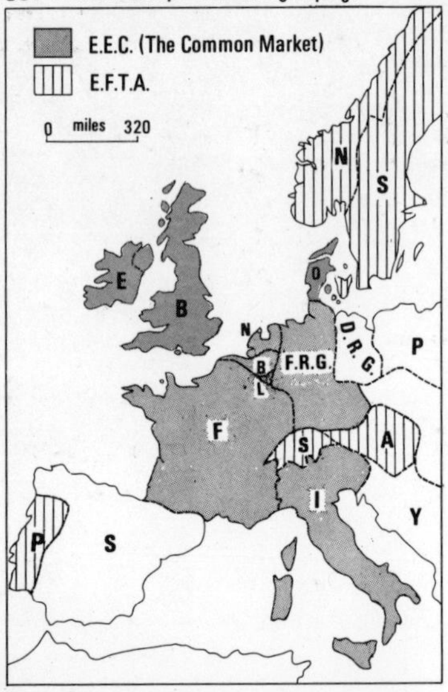

388 Population – Britain and the E.E.C. countries

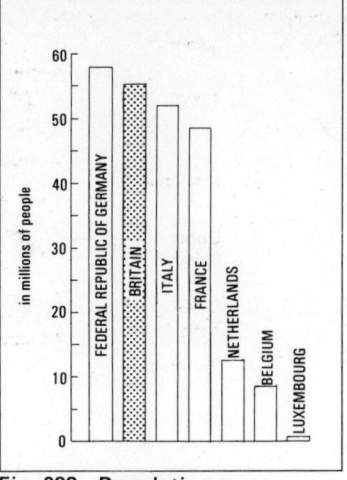

389(i) Britain and the E.E.C. – Crude steel production

389(ii) Britain and the E.E.C. – Coal and Lignite (Brown coal) production

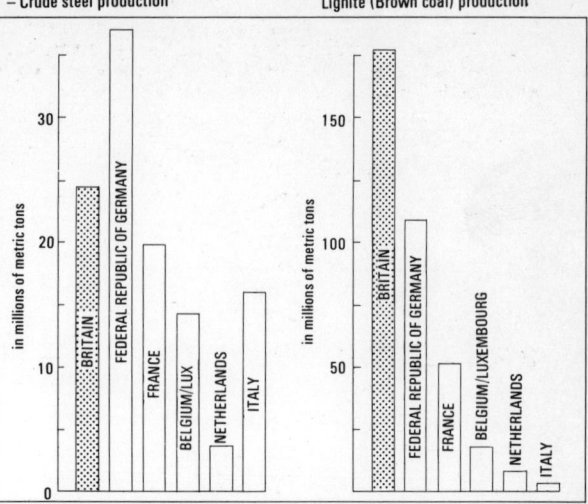

Fig. 387 Economic groupings

387a (i) For what do the initials E.E.C. and E.F.T.A. stand? (See Fig. 392.)
(ii) To which economic group does Britain belong?

387b (i) List the countries in each group.
(ii) Which group is more compact?

390 Britain and the E.E.C. – Motor vehicle production

Total production 8,525,000 vehicles

Fig. 388 Population

388a How does the population of Britain compare with that of other E.E.C. countries?

391 Europe – Main centres of population

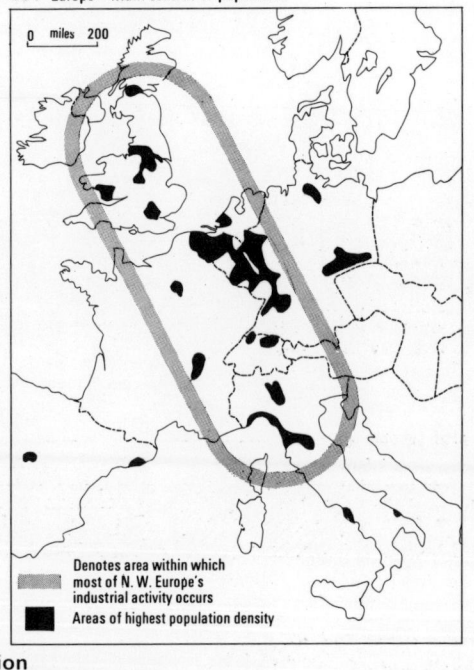

Denotes area within which most of N.W. Europe's industrial activity occurs

Areas of highest population density

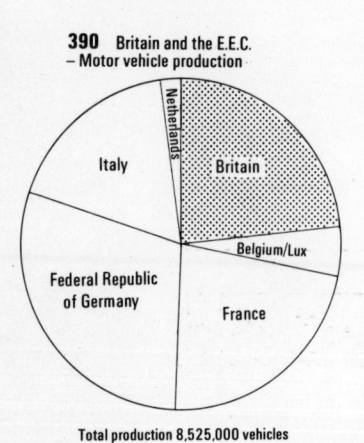

Fig. 390 Motor vehicle production

Most Western European countries rely on valuable exports of manufactured goods to allow them to import the large quantities of raw materials and foodstuffs they cannot provide themselves. The motor vehicle is one of the most important exports.

390a (i) Which are the four most important vehicle-producing countries in Western Europe?
(ii) Name one car firm in each.

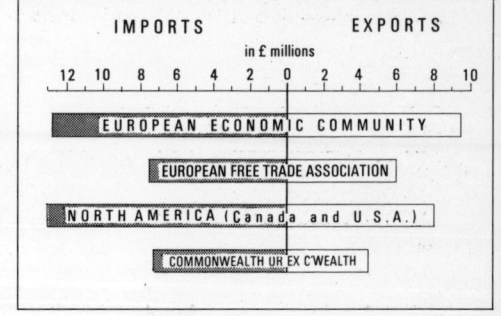

Figs. 392, 393 and 394   World and European trade

392a Which world region or economic group does Britain trade with most?

393a Is Britain's balance of trade with other E.E.C. countries in surplus or deficit?

393b (i) What are the main types of goods passing between Britain and other E.E.C. countries.
(ii) Explain why.

394a With which three E.E.C. countries does Britain do most trade?

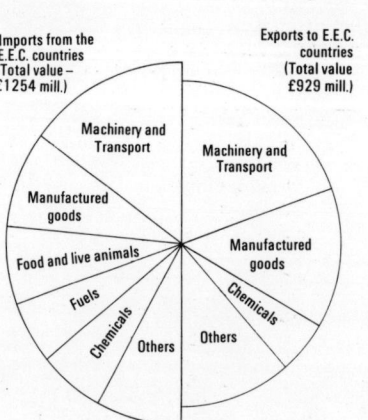

Fig. 389 Coal and steel production

389a (i) Which country is the largest crude steel producer in the E.E.C.?
(ii) How does its output compare with that of Britain?

389b Explain why steel is a significant indicator of a country's industrial importance.

389c (i) What country is the most important coal producer in North-West Europe?
(ii) Why is coal production decreasing?

Fig. 391 Population distribution

Large urban areas grow up where industry develops.

391a (i) Indicate the main urban areas in Western Europe.
(ii) Find out the main industries in each.

392 Britain – Trade with selected world regions and groupings

IMPORTS   EXPORTS
in £ millions

EUROPEAN ECONOMIC COMMUNITY
EUROPEAN FREE TRADE ASSOCIATION
NORTH AMERICA (Canada and U.S.A.)
COMMONWEALTH UK EX C'WEALTH

Imports from the E.E.C. countries (Total value – £1254 mill.)

Exports to E.E.C. countries (Total value £929 mill.)

393 Britain and the E.E.C. – Trade

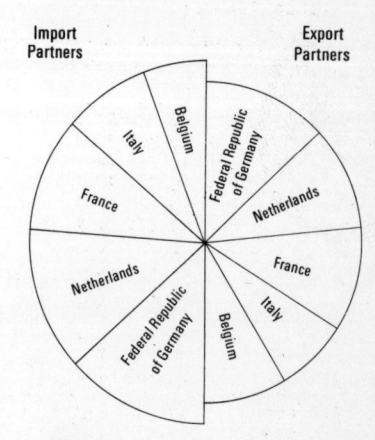

394 Britain and the E.E.C. – Trading partners